D0397000

Hillsboro Public Library
Hillsboro, OR
A member of Washington County
COOPERATIVE LIBRARY SERVICES

PRAISE FOR
THE HEALING ORGANIZATION

"*The Healing Organization* is based on a simple, profound revelation: Businesses that operate from love and make human flourishing their first priority will enrich the lives of all their stakeholders, generate more sustainable abundance, and can help solve many of our current crises. Raj Sisodia and Michael J. Gelb bring this world-changing idea to life with deeply moving stories of awakened consciousness."

—Deepak Chopra, MD, author of *Metahuman*

"This book is a vital contribution and will help accelerate the creation of a truly human future. Raj Sisodia and Michael J. Gelb have authored a powerful consciousness-raising guide based on the new paradigm of balance between feminine and masculine energies. When these energies are in harmony, we all experience healing—personally, organizationally, and societally. *The Healing Organization* is essential reading for all aspiring and highly conscious leaders!"

—Kristin Engvig, founder & CEO, WIN and WINConference; winner of the 2010 International Alliance of Women (IAW) Award

"A great—and important—book! At its heart, Raj Sisodia and Michael J. Gelb's *The Healing Organization* is a deeply spiritual book that shows leaders the way forward to transform business to be the greatest healer of societal ills, while creating wealth for all its stakeholders. If you follow its wise advice, you will build organizations that make this world a better place for all to live in."

—Bill George, Senior Fellow, Harvard Business School; former chair & CEO, Medtronic; author of *Discover Your True North*

"*The Healing Organization* is a mind-blowing, transformational guide that shows how business can lead the way to alleviate suffering and elevate joy. Raj Sisodia and Michael J. Gelb bring a rare combination of scholarly insight and practical guidance to this compelling tour de force. This is essential reading for all aspiring leaders."

—Marshall Goldsmith, PhD, #1 *New York Times* bestselling author of *Triggers, MOJO,* and *What Got You Here Won't Get You There*

"A must-read for all leaders, and especially politicians, at every level! This urgently needed book presents a road map for bringing left and right together by transforming the way we think about the relationship between expanding prosperity and social good. *The Healing Organization* heralds a new epoch, a powerful idea for a better world whose time is now."

—Kerry Healey, PhD, president of Babson College; former lieutenant governor of Massachusetts

"If you want to do well by doing good, if you want to align your business with your higher purpose, if you want to be a leader who makes a truly positive contribution to your people and your community, then this book is for you! *The Healing Organization* is a game changer. Raj and Michael are leading the way to the next evolution of conscious capitalism."

—John Hope Bryant, CEO, Operation HOPE, Bryant Group Ventures, and The Promise Homes Company

"In this beautifully written and urgently needed book, Raj Sisodia and Michael J. Gelb present the case for a radical rethinking of the purpose and possibilities of business. They do this first with a succinct and incisive analysis of the evolution of contemporary capitalism. The authors illuminate the true genius of Adam Smith, guiding us to reconsider *The Wealth of Nations* in the light of Smith's *Theory of Moral Sentiments*. Then they share a series of heart-opening, mind-expanding, and profoundly inspiring stories that show what's possible when the heroic spirit of business is liberated through this synthesis. Finally, they present a guide, a playbook, and an Oath for aspiring Healing Leaders. Contemplate the guide. Follow the playbook. Take the Oath. Help make the world a better place."

—John Mackey, cofounder & CEO, Whole Foods Market; coauthor of *Conscious Capitalism: Liberating the Heroic Spirit of Business*

"*The Healing Organization* is a deeply moving, inspiring, and important work. It builds the historical, philosophical, and most importantly, the human, emotional case for the next natural evolution of capitalism.

Gelb and Sisodia present BIG IDEAS in a compelling, illuminating, and engaging way. More than a business book, it's a manifesto that offers hope for a new vision that can heal the divisiveness in our society.

And, as a business book it is a 'must read,' especially for the emerging generation of entrepreneurs I serve as investor, strategic advisor, and mentor. They want to generate wealth in a way that makes a positive difference in the world. They want to live with a sense of meaning and purpose. They want to enrich the lives of all their stakeholders. *The Healing Organization* shows how this can be done, how it is being done and why, for the sake of humanity, it must be done, now!"

—Jonathan Miller, CEO, Integrated Media Co.; former CEO, AOL

"Sisodia and Gelb are master storytellers, and this amazing book celebrates the potential of business to be a powerful force for good in our world. It's inspiration at its best to read of companies such as Danny Meyer's Union Square Hospitality Group transforming the lives of their employees and other stakeholders, while generating more profit than less enlightened businesses. This book brought tears to my eyes, new ideas to my mind, and hope to my heart—it is a must read for every business school aspirant and graduate."

—Karen Page, MBA, award-winning author of *The Flavor Bible* and *What to Drink with What You Eat*

"A beautiful work with a powerful message, much needed for our times, illuminating the flowering of progress in a messy world. Reminds one of the Buddha's observation upon seeing a lotus growing from the mud: 'Om Mani Padme Hum.' This book is filled with insights for creating a more enlightened society based on profitable businesses with more sustainable and caring values. It is a compelling work that inspires one to change and evangelize the message."

—Ketan Patel, author of *The Master Strategist*; founder & CEO, Greater Pacific Capital

"In an imperfect world, what is the role of business? Across the globe, political leaders, corporate shareholders, and ordinary citizens are reckoning with that question. And it's one for which Sisodia and Gelb offer a provocative answer: business can—and should—help heal society's ills. This smart and meaningful book will inspire you to lead your organization into a place of joy, health, and yes, profitability, that can deliver greater well-being to employees, communities, and the planet."

—Daniel H. Pink, author of *WHEN* and *DRIVE*

"Climate change and gross inequality now threaten the natural world and the future of humanity. With time running out before we reach irreversible tipping points, business must take the lead in healing our world by investing in our most important assets: our natural and human capital. This book provides a compelling case for why this must happen now and more importantly, it offers the practical steps to show how this can be accomplished. Highly recommended!"

—Paul Polman, former CEO, Unilever

"In this wonderful book, Michael J. Gelb and Raj Sisodia illuminate the historical, psychological, and philosophical underpinnings of contemporary capitalism to make a compelling case for reimagining the fundamental purpose of business. Business is the straight line between where we are now as a society and making the world a better place. If you want to improve the world, then please read this book and get it into the hands of every influential person you know."

—Kip Tindell, cofounder and chairman, The Container Store; cochairman of Conscious Capitalism, Inc.

"This brilliant book tells a new, emerging story of money and business based on much wiser assumptions about human possibility. The authors show how businesses with a higher purpose, basic goodness, and concern for the welfare of all stakeholders, including communities and the Earth itself, are thriving. Instead of causing or exacerbating suffering, these businesses prevent and alleviate it. By meeting real needs and operating from love rather than fear, these companies elevate joy and generate superior financial returns. This is an inspiring, soul-nourishing, and practical guide to transforming the role of business in society for the betterment of all."

—Lynne Twist, author of *The Soul of Money*; cofounder, The Pachamama Alliance

THE
HEALING
ORGANIZATION

Raj Sisodia
and Michael J. Gelb

THE
HEALING
ORGANIZATION

Awakening the
Conscience of Business
to Help Save the World

HarperCollins
Leadership

An Imprint of HarperCollins

© 2019 Raj Sisodia and Michael J. Gelb

All rights reserved. No portion of this book may be reproduced, stored in a retrieval system, or transmitted in any form or by any means—electronic, mechanical, photocopy, recording, scanning, or other—except for brief quotations in critical reviews or articles, without the prior written permission of the publisher.

Published by HarperCollins Leadership, an imprint of HarperCollins Focus LLC.

Book design by Aubrey Khan, Neuwirth & Associates.

ISBN 978-0-8144-3982-1 (eBook)
ISBN 978-0-8144-3981-4 (HC)

33614081503582

Library of Congress Cataloguing-in-Publication Data

Library of Congress Control Number: 2019936544

Printed in the United States of America

19 20 21 22 LSC 10 9 8 7 6 5 4 3 2 1

IN MEMORIAM

We commemorate and celebrate the lives of two extraordinary leaders who inspired the ideal of the Healing Organization.

Herb Kelleher (1931–2019) taught us "the business of business is people—yesterday, today, and forever." Herb was wise and playful, tough-minded and tenderhearted: a man in full, a whole human being, a beacon for us all.

Bernie Glassman (1939–2018) taught us to bear witness to suffering and to creatively use the power of business to uplift human beings and offer those dealt a tough hand in life a "first chance."

DEDICATION

Raj: With love and gratitude to Zahir Ahmed Quraeshi. Over the past thirty-four years, you have continually inspired me with your dignity, grace, and commitment to duty. Your warmth, irrepressible humor, lightness of being, and open heart make you a healing presence in the lives of all who are blessed to know you.

Michael: With love and gratitude to Deborah Domanski. This book is about Truth, Beauty, and Goodness and you bring those qualities to my life every day. Listening to your exquisite voice as I wrote elevated my spirit and inspired my creativity. The world is blessed and healed by your artistry, grace, humor, intelligence, and loving-kindness.

CONTENTS

CONTENTS

FOREWORD

BY TOM PETERS

I am the luckiest guy in the world. From the moment it's released, *The Healing Organization* is going to be a classic (I really dislike that word, "classic," but upon rare occasion it's merited). Within a few years, I'll further guarantee (another overused word) that this book will appear on all-time "best business books" lists. It is an extraordinary book, and I get to write the Foreword—hence, my name will be directly associated with this masterpiece. As I said, lucky me.

The Healing Organization defines the word "audacious." Consider this opening promise: "You will discover how business can become a place of healing for employees and their families, a source of healing for customers, communities and ecosystems, and a force for healing in society . . ." Audacious, right?

Well, I buy the act, no matter how audacious. The best news of all: While the authors meticulously and extensively build their intellectual foundation, the heart of the book is cases, cases, and more cases.

The title of Part 2: "The Joy That Is Possible: Stories of Organizations That Heal." First up: "The Power of Innocence: How Jaipur Rugs Brought Dignity, Prosperity, and Hope to 40,000 Lower Caste Women in India." The title evokes that word—*audacious*—again. But by the time you finish the Jaipur Rugs saga, I am almost certain that you will conclude, as I did, that there is not a shred of hyperbole in the bold chapter title. From Jaipur, India to Yonkers, New York: "The Zen of Brownies: How Greystone Bakery's Open Hiring Policy Is Healing the Cycle of Crime and Incarceration." (FYI: In case you wondered, Greystone bakes 35,000 pounds of brownies every day.)

And Eileen Fisher: "The Most Interesting Woman in the World: How Eileen Fisher Empowers Women to Move Freely and Be True to Themselves." And the Motley Fool: "Where are the Customers' Yachts? How Fools, a Shaman, and Larry Fink Are Beginning to Heal Wall Street." And on it goes.

All of this is personal to me.

Thirty-seven years ago, in 1982, I coauthored a book titled *In Search of Excellence*. The heart of the book was . . . cases, cases, and more cases. At the top of the list of eight "basics" Bob Waterman and I teased out of the data was . . . putting people first. Thirty-six years after that, in 2018, I published my seventeenth book, *The Excellence Dividend*, and the main message was . . . putting people first. I continue to be amazed that "people first" is still news. To publicize the new book, I appeared in dozens of podcasts, and I swear to you that the opening line from the interviewer was almost always word-for-word the same: "Tom, you write a lot about people; tell us about that . . ." And my answer in effect was always the same: "WHAT THE HELL ELSE IS THERE?" (Stronger language coursed through my head.)

Well, alas, we know what else there is. Shareholder value maximization has been the most devastating idea in modern business history. In a 2014 *Harvard Business Review* article, "Profits Without Prosperity," Professor William Lazonick reported on his assessment of 449 S&P 500 companies. Over a ten-year period, 91 percent of profits went to share buybacks and dividends, leaving a paltry 9 percent for "productive capabilities or higher incomes for employees." The tide may be, as some are saying, turning, but we are a long, long way from "people first," let alone "the Healing Organization."

But we keep trying.

As Margaret Thatcher famously said, "It's a funny old world." The day after finishing this foreword, I will head to Ann Arbor, Michigan, to spend the day with Rich Sheridan, the CEO of the software company Menlo Innovations, one of the Healing Organizations featured in this book. Rich, author of *Joy, Inc.*, explains: "It may sound radical, unconventional, and bordering on being a crazy business idea. However—as ridiculous as it sounds—joy is the core belief of our workplace. Joy is the reason my company, Menlo Innovations, a customer software design and development firm in Ann Arbor, Michigan, exists." Amen.

Rich and I will be discussing plans pushing People First/Joy further into the light. I intend to fight the good fight until my last breath. And now, the arrival of *The Healing Organization* dovetails perfectly with my effort.

There is a special reason, circa 2019, to put more fuel on the fire. The "tech tsunami," as some call it, may well wreak havoc on employment in the next one to three decades. To me, that means there is what I call A New Moral Imperative that moves "people first" (*Joy, Inc./The Healing Organization*) to the forefront of societal consciousness. For one—very big—thing, there can be little doubt that the rise of populist movements around the globe is in part driven by anticipation of massive employment dislocations. In *The Excellence Dividend*, I offered what I labeled Corporate Mandate #1: "Your principal moral obligation as a leader is to develop the skillset of every one of the people in your charge (temporary as well as semi-permanent) to the maximum extent of your abilities and consistent with their 'revolutionary' needs in the years ahead. The bonus: This is also the #1 profit maximization strategy!"

People first.

Moral and societal imperative.

Business success.

Joy, Inc.

The Healing Organization.

Now much, much more than ever.

This book constitutes no less than a magnificent contribution to the life-and-death process of reimagining organizations and society itself. *The Healing Organization* is indeed audacious. But the facts are here. The cases are here. And God knows, the need is here.

Read.

Ingest.

Share.

Reimagine.

Act.

Act with audacity.

Now.

Tom Peters
South Dartmouth, Massachusetts
May 12, 2019

PROLOGUE

A SACRED UNDERTAKING

FROM RAJ

After finishing high school in India with good grades in math and science, I did what was expected and went to engineering school, despite having no passion for engineering. After graduating I went on to business school because I was told my salary would double and I could work in an air-conditioned office. I then went to Columbia University to do a PhD in marketing and business policy—primarily so I could return to the US, where I had spent a formative part of my childhood.

While intellectually stimulating, I found the experience of studying and then teaching business at odds with my own trusting, idealistic, and peaceful sensibilities. I never felt resonant with the dominant view that business had to be a "dog-eat-dog" world in which "only the paranoid survive."

Coming from India, then a commercial backwater, I was stunned at the sheer omnipresence of marketing in US culture. I found a lot of it unethical, much of it wasteful, and most of it ineffective. My academic work over the succeeding two decades focused primarily on describing what was wrong with marketing. My colleagues and I showed that spending had gone up dramatically while customer loyalty and trust had plummeted; only 8 percent of Americans had a positive view of marketing.[1] In 2004, it was estimated that companies spent $1 trillion on marketing, which was equal to the GDP of India that year.[2] Today that number has increased further: Americans each received approximately 41 pounds of junk mail in 2016, or 6.7 billion tons collectively—most of which is never recycled.[3] I wondered: What were we getting from this tsunami of spending? How was it benefiting customers, companies, and society? My conclusion was that marketing was doing more harm than good.

I was going to alert society by publishing these troubling statistics and more in a book entitled *The Shame of Marketing*. Fortunately, my mentor Jag Sheth gave me a sage piece of advice. He said, "Raj, in America, people would rather hear about the solution than the problem."

That simple insight turned my life around. I relabeled the book *In Search of Marketing Excellence*, and started to look for companies that spent modestly on marketing and yet had outstanding customer loyalty and trust. That book evolved into *Firms of Endearment: How World-Class Companies Profit from Passion and Purpose*, and it led to the identification of the four pillars of what we would later call Conscious Capitalism.

I remember the moment when I found my purpose—or, more accurately, when my purpose found me. On June 12, 2005, I was researching stories of how some companies were demonstrating deep and authentic caring about their customers, employees, and communities and found myself moved to tears. I had never had a *positive* emotional response to my work before. I realized "There's a better way." Not only did I find one, I discovered a bonus to the better way: *It wasn't just more caring and humane, it was also far more profitable.*

I gradually began to understand that business could help to heal, instead of contributing to the suffering I saw all around the world. I saw the suffering near my mother's village, where the beautiful, life-giving Chambal River had been reduced to a poisonous trickle by effluents from a textile plant. I saw it in my father's village, where most birds have disappeared, as have the butterflies, bees, and earthworms, because of an overreliance on insecticides and pesticides. I saw it in startling statistics about the pervasiveness of financial distress in the US, about rising rates of anxiety, depression, and suicide, and about increasing cultural discord. I viscerally felt the need for healing at every level, from within my own being to the world at large.

I was energized and motivated to learn as much as I could about organizations that had a healing effect on their stakeholders. When I thought about it, I experienced chills and goose bumps—a sure sign that this was an idea I needed to pursue. As I continued, people with compelling stories of healing organizations kept showing up!

Wonderful collaborators have appeared via synchronicity for all my books. For this book, I knew that there was only one possible coauthor: Michael Gelb. Michael had come into my life when I was thirty-nine years old, right when most people experience the much-joked-about but very real midlife crisis. His presence, energy, and wisdom changed my life. Michael showed me that I could be a creative and whole person, not just the left-brained, hyper-analytical type that I had pigeonholed myself to be. He helped me gain the courage to trust my own instincts and listen to my inner voice, so that I could eventually discover and fulfill my own unique purpose.

As we started to work on this book, Michael said something that had a profound impact on me: "Writing this book is a sacred undertaking. We have to do justice to the challenge we have chosen." We have both carried this commitment with us throughout the journey. Every word has been filtered through our souls.

We have written this book with love, joy, and a deep sense of responsibility. Having completed the necessary research and inner work, we also write now with a sense of urgency. There is no time to waste; it is later than we realize.

My vision—symbolically, but also practically—is to have the other side of the river near my mother's village return to being green and blue again, to have the water restored to its pristine beauty, and to return to my father's village and hear the birds sing once again.

FROM MICHAEL

When I graduated from Clark University with a double major in psychology and philosophy, I set criteria for what I wanted to do with my life: It had to be something healing for others and for myself. This led me to spend a year studying the world's wisdom traditions and meditation practices with J. G. Bennett in England. In 1978, I completed a three-year certification training as a teacher of the Alexander Technique of Mind/Body Coordination. In the same year I received my master's degree from Goddard College and soon thereafter my thesis was published as my first book: *BodyLearning*. During this time I met and began collaborating with the originator of Mind Mapping, creative thinking pioneer Tony Buzan. Together, we developed and led five-day "Mind & Body" seminars for senior corporate leaders globally. In 1982, I was certified as the first Master Trainer of Buzan's work. Later that year, with an idealistic dream to help save the world, I moved to Washington, DC—a place where it seemed that creative thinking, accelerated learning, and innovative leadership strategies were most desperately needed.

I began offering open-enrollment three-day High Performance Learning seminars, but was disappointed to discover that there were only a few registrants from government or the political sphere. Fortunately, the programs were popular with businesspeople and this led to many opportunities for me to teach and consult with companies in the DC area and beyond. This was the beginning of my realization that the dynamism of business made it, rather than government, the greatest point of leverage for making a positive difference in the world.

In 1997, I was asked to lead a series of these seminars as part of the Executive MBA program at George Mason University, to help participants learn the skills they needed to lead innovation efforts at work. Raj Sisodia, the director of the program who invited me to conduct the classes, was genuinely curious and open. He had a wonderful passion to enrich the lives of his students. I enjoyed our collaboration and we became friends.

Then in 2006, Raj sent me a copy of the draft manuscript of his seminal book *Firms of Endearment: How World-Class Companies Profit from Passion and Purpose.* Raj and his coauthors made a compelling business and academic case for what I had dreamed might be possible. Suddenly, I realized that I wasn't just a solo practitioner with a quixotic notion of making a better world through helping businesses become more creative, conscious, and compassionate; I was part of a movement. With the support of John Mackey of Whole Foods Market, Kip Tindell of The Container Store, Doug Rauch of Trader Joe's, and many others, this movement became Conscious Capitalism.[4] I was thrilled when Raj invited me a few years later to keynote the annual Conscious Capitalism conference and then to serve as master of ceremonies for the Conscious Capitalism CEO Summit.

I shared with Raj how much his books had inspired me and he said that my books had a similar effect on him. So it was natural for us to explore the possibility of writing something together. We both knew business leaders who were changing the world by creating positive, people-centered organizations and it was clear that they were having a healing effect on all their stakeholders. What if we explored further how and why they did this? What if we shared their stories to help more people realize and be inspired by what is possible? And what if these Healing Organizations were the key to mitigating many seemingly intractable problems like environmental degradation and climate change, obesity, opioid addiction, rising rates of anxiety, depression and suicide, and even the gaps between rich and poor, left and right?

For forty years, I have worked with visionary leaders around the world to support them in nurturing more innovative and human-centered cultures and to equip them with creative thinking tools and strategies that help translate ideals into reality. *The Healing Organization* represents an expansion of my own learning about what's possible. As we worked on each story, I found myself moved to tears by the courage, tenacity, and sheer goodness that emanates from each of them. Tempered by the decades, the dream with which

I began my career is stronger and more vital than ever: I dream that together we can create a new story of business based on awakened conscience, through which we can help save the world.

FROM MICHAEL AND RAJ

Writing this book is, for us, a sacred undertaking.

It is something that we *had* to do. It feels like our entire lives have been building toward this project.

Our process of writing has been a joyful experience of collaborative learning and discovery. Inspired by David Cooperrider's *Appreciative Inquiry,* we suspended our preconceptions and opened our minds and hearts to learn as much as possible from each series of interviews. Wherever possible, we visited the companies and met the leaders in person. There are many other Healing Organizations that could be featured in this book. The ones we profile here are those we know best.

The book is written in a unified voice. The *we* that is behind the words that follow is an expression of our shared healing purpose.

We want to alleviate unnecessary suffering—physically, emotionally, spiritually, and financially—caused by the way most business is done. We seek nothing less than the transformation of the workplace from a place of stress and fear to one of inspiration and growth, from what feels to many like a miserable prison to a joyful playground.

This book is not about the business of healing; it is about business *as* healing.

The scope of our concerns goes beyond employees and their families and includes all those whose lives are touched by the company: customers, suppliers, communities, citizens. Crucially, it includes the environment and all life on the planet.

And, now, as you read and contemplate what follows, we hope that it will include you.

HOW TO
GET THE MOST
FROM THIS BOOK

There is a logical, linear order to this book, but following it is optional. We invite you to read it in a way that best suits your needs and curiosity.

Perhaps, like many of our friends, you are intrigued by the possibility that business can be a source of healing and abundance for all its stakeholders but you're skeptical about whether that's really possible. If so, you may wish to begin by going straight to part 2, where you will be inspired and amazed by the compelling, uplifting stories of companies that are living this ideal.

Our first story, about Jaipur Rugs, features a company that transformed its industry and the lives of 40,000 women and their families, while making consistently superior profits. If this can happen in an impoverished rural area in India, then imagine what can be accomplished in the United States and other prosperous societies.

If you work in an industry or organization that is not healing, then you may wish to explore the stories of companies that underwent profound transformations from hurting to healing such as DTE Energy, Appletree Answers, and FIFCO.

If you are already part of a company with a conscious, positive culture but you want new ideas and inspiration on how to evolve further, then you may wish to start with stories of companies that have discovered innovative ways to heal their stakeholders, such as Menlo Innovations, Union Square Hospitality Group, and KIND Snacks.

Maybe you're already familiar with and inspired by some of these Healing Organizations and you want to understand the principles and practices that they all have in common? In that case, you may want to begin with part 3. You can even start with the epilogue and begin your journey through this book by taking the Healing Oath on the last page.

If you are curious about the historical, psychological, and philosophical underpinnings of capitalism and you want to understand how we have reached our current inflection point—where business can and must take the lead in helping to save the world—then you'll want to begin with part 1.

This isn't a book about corporate social responsibility or why it's nice to have a wellness program, and it isn't about checking a few boxes on environmental stewardship or finding more humane tactics to squeeze out more profit; rather, it's about a mind-blowing, heart-opening, world-changing rethinking of business. We make the big-picture case for this new paradigm in part 1.

Wherever you begin, and in whatever order you read, you'll get the most from the book if you keep a journal or notebook and jot down the ideas as they inspire you. You will also benefit from sharing the ideas, insights, and examples with friends and colleagues, as they offer a desperately needed dose of hope and optimism. As you share this hope and optimism, these qualities will strengthen within you.

INTRODUCTION

TRANSFORMING THE WORKPLACE FROM PRISON TO PLAYGROUND: THE CHOICE IS OURS

What if there was a company whose CEO said, "Let's pay our people as much as possible"? What if that same company showed up to help communities in distress before FEMA and the Red Cross when there was a disaster? What if your company treated your spouse, your children, your parents, and even your pets, as stakeholders, and made it a priority to ensure that you can be present for them? What if there was a company that not only reduced its environmental footprint to zero, but actually made a net positive contribution to its ecosystem? What if there are companies that gladly hire some of the 70 million Americans with criminal records and give them what most of them never had—a *first* chance, an opportunity to build a happy life, raise a family, and become productive, taxpaying citizens?

And what if these businesses, which prioritize the welfare of all their stakeholders, and help heal their employees, customers, and communities, *are more profitable and prosperous than their industry peers*?

Such businesses, and many others like them, do exist, and we share their inspiring stories in this book. We emphasize that these businesses are more profitable because we believe that profit is a social good. It is socially irresponsible for a business to not be profitable; a free society cannot function without profitable businesses. The quest for wealth fuels creativity, innovation, and entrepreneurship; moreover, without profits, there is no tax revenue, and without tax revenue there can be no infrastructure or public services. But, as the companies and leaders profiled here all understand: *It matters how you make the money.*

You will discover how business can become a *place* of healing for employees and their families, a *source* of healing for customers, communities, and ecosystems, and a *force* for healing in society, helping alleviate cultural, economic, and political divides.

A TALE OF TWO PARADIGMS

Business pervades our lives. More than governments, nonprofits, or religious institutions, business is the dominant force in contemporary life, for better and for worse. In free societies, the vast majority of our needs are fulfilled by corporations and small businesses. Most people are employed by private enterprises. The ways in which these organizations operate has a huge impact on every aspect of our lives: our material well-being; our physical, emotional, mental, and spiritual health; and our ability to be present and function well as parents, spouses, community members, and citizens.

For the most part, businesses have succeeded in meeting our material needs and elevating our collective flourishing. Over the past two hundred years, almost every indicator of human well-being has risen sharply in concert with the spread of free market capitalism. We live longer, are more educated, produce and consume more, and enjoy more leisure time than our ancestors. We have access to miraculous technologies that allow us to access people and learn nearly anything in just a few keystrokes. And we are living in the most peaceful and prosperous time in history. This is all enabled by the dynamism and innovation that are the hallmarks of the capitalist system.

Many people are thriving. There's more freedom for entrepreneurs to generate wealth in creative and useful ways. As Greek poet Homer observed 2,500 years ago, "Each man delights in the work that suits him best."[1] The opportunity to discover and pursue our ideal work is greater than ever before.

Yet, even as we progress in so many ways, we allow drastic, unnecessary suffering to continue—and business plays a huge role in causing it. The pandemics of obesity, opioid addiction, alcoholism, depression, anxiety, gun violence, and the corruption of our planetary ecosystem are all exacerbated by the way business is conducted.

In *A Tale of Two Cities*, Charles Dickens sums up the paradox in which his characters lived and in which we live today:

> It was the best of times, it was the worst of times, it was the age of wisdom, it was the age of foolishness . . . it was the spring of hope, it was the winter of despair, we had everything before us, we had nothing before us, we were all going direct to Heaven, we were all going direct the other way.[2]

The direction we go depends on how we evolve our thinking about business. If we continue on our current path, then foolishness and despair will prevail. If we evolve and transform, by applying wisdom and light to change how we think about organizations, heaven awaits.

A more contemporary Homer—Homer Simpson—mused: "Alcohol: The cause of and solution to all of life's problems."[3] Good for a laugh but off target in his diagnosis and prescription. The cause, and the solution, to many of the world's problems is *capitalism*—specifically, the way we think about it and practice it.

ESCAPING THE PRISON OF OUR THOUGHTS

Matt Groening, creator of *The Simpsons*, quipped, "I'm not going to stop torturing myself until I figure out the cause of my pain."[4] What is the cause of our pain, and how much of a role does capitalism play in it?

At work, many people feel that suffering is inflicted upon them by their managers, bosses, and supervisors. Dr. Edith Eva Eger, author of *The Choice*, who survived the Auschwitz death camp, explains, "We are victims of victims."[5] In other words, those who torment us, whether schoolyard bullies or bad bosses, were most likely themselves tormented by someone else.

We can't always control our circumstances, but as Matt Groening reminds us, much of our suffering is self-inflicted. It happens because we don't know how to think about life and how to manage and regulate our own emotions. Attitude matters profoundly, on an individual and a collective basis. When the collective attitude accepts and even embraces the idea that it's a "dog-eat-dog" world and that business is nothing more than a rat race, that work and life are separate entities that need to be balanced instead of integrated, much suffering results.

As individuals, we can learn to change our patterns of thinking, our attitudes, so that we don't torture ourselves, and we can change the way we think about business so we don't torture one another.

The need for this change is acute because circumstances are still extremely harsh for the majority of people. Although global life expectancy has risen dramatically, fifteen thousand children under the age of five still die every day from preventable causes.[6] And, although, according to data compiled by the World Bank, more than one billion people have been lifted out of extreme poverty (defined as $1.90 per person per day) since 1990, half the world's population still live on less than $5.50 a day. Life for nearly half of humanity remains a daily struggle for survival.[7]

But suffering doesn't disappear with rising prosperity. For example, the towns of Aspen, Colorado, and Palo Alto, California, are two of the most prosperous communities in the United States, indeed in the world. Yet, they also have rates of depression, addiction, and suicide that far exceed the US average. Aspen's rate is three times the US average, while Palo Alto has the highest teen suicide rate in the country.[8] Whatever one's level of income, when meaning and purpose are absent, when we feel dehumanized and objectified, we experience emotional and spiritual suffering.

The objectification of humans by business was partly a consequence of the materialist paradigm of the modern scientific era that started about three hundred years ago. It created a view of the universe as a clockwork machine in which separate objects behave in predictable ways based upon fixed laws in time and space. This view has been overturned by contemporary science, which validates Leonardo da Vinci's observation that "everything is connected to everything else."[9] As Einstein expressed it:

A human being is part of the whole, called by us "Universe"; a part limited in time and space. He experiences himself, his thoughts and feelings as something separated from the rest—a kind of optical delusion of his consciousness. This delusion is a kind of prison for us, restricting us to our personal desires and affection for a few persons nearest us. Our task must be to free ourselves from this prison by widening our circle of compassion to embrace all living creatures and the whole of nature in its beauty.[10]

Contemporary organizational structures are even more influenced by misinterpretations of evolutionary biology than they are by outdated conceptions of physics. Misinterpretations of Darwin have generated a narrative about human society that is rooted in a zero-sum, survivalist mindset. In his classic work *The Descent of Man*, Charles Darwin only mentioned the phrase "survival of the fittest" twice; he mentioned the word "love" ninety-five times.[11] Darwin believed that empathy is our strongest instinct and the reason we human beings have succeeded the way we have. In a sense, we *are* born to be our brother's keepers; cooperation is essential to our thriving as a species.

The mindset that humans are motivated primarily by feelings of scarcity, separation, and competition formed the basis of behavioral psychology and the dominant views of contemporary economics. Economist, banker, and social entrepreneur Muhamad Yunus was awarded a Nobel Prize for challenging these assumptions. He states, "Indifference to other human beings is deeply embedded in the conceptual framework of economics. . . . We need to design a theory keeping in mind the true human being, not a distorted and miniaturized version. A true human being is selfless, caring, sharing, trusting, community-building, friendly—and at the same time, the reverse of all these virtues."[12] In other words, how do we design organizations and organizational systems that uplift the "true human being"? How do we create companies that promote our selfless, pro-social nature while liberating us from distortion and miniaturization?

As author Louis Menand, winner of the National Humanities Award, writes, "We invented our social arrangements; we can alter them when they are working against us. There are no gods out there to strike us dead if we do."[13]

DYING FOR A PAYCHECK

This reformulation is urgent because work as it is currently constituted is striking too many people dead.

The Japanese (*karoshi*) and the Chinese (*guolaosi*) have had to create words for "death from overwork." An estimated 10,000 Japanese die from overwork every year. According to the *China Youth Daily*, 600,000 people a year die in China from working too hard.[14]

Most people are relatively healthy and whole when they start their work lives. But over time, the stress of the workplace wears them down and they develop chronic health conditions. Heart attack rates are highest on Monday mornings, when people literally no longer have the heart to get up and go to work for another soul-destroying week.[15]

More than wars, murderers, and terrorists, our work is killing us. The human cost of "business as usual" has become unacceptably high. We are sophisticated at cost accounting; we know every expense down to the penny, but we don't even *try* to measure human suffering. Instead we just say, "Thank God it's Friday," because work for many of us is an ordeal to be survived.

In his landmark 2018 book, *Dying for a Paycheck: How Modern Management Harms Employee Health and Company Performance—and What We Can Do About It,* Stanford professor Jeffrey Pfeffer emphasizes that the most dangerous occupations used to be in industries like coal mining and oil exploration or chemical manufacturing. But physical dangers from work have been greatly reduced, thanks to the efforts of agencies like the Occupational Safety and Health Administration (OSHA). Today, white-collar work has become far more hazardous to our well-being than blue-collar work used to be. Stress-related conditions account for more than 75 percent of visits to doctor's offices, and work pressures and financial worry are by far the greatest stressors in contemporary life. Pfeffer and his colleagues estimate that stressful workplace environments are responsible for at least 120,000 excess employee deaths each year in the US alone, and $180 billion of additional health-care expenditures.[16]

Many times that number are suffering in ways that fall short of dying but are still devastating. Many businesses also contribute to suffering and ill health for the families of their employees, for customers, and for citizens generally through pollution. They are also involved in the systematic abuse and inhumane treatment of animals in the factory farming system.

All suffering matters, including that of animals. Whether you consider yourself religious, spiritual, an atheist or somewhere in between, has it ever occurred to you that the way you treat other living beings, including animals, impacts us all? How are your actions adding to or taking away from the suffering in our world?

WHAT DO WE MEAN BY CONSCIENCE?

"Conscience," states pioneering educational genius Horace Mann, "is the magnet of the soul."[17] George Washington called it the "spark of celestial fire" and urged that we keep it alive in our hearts.[18] The word is derived from the Latin *conscientia* meaning "knowledge within oneself." Most people have the knowledge within themselves that life is interconnected, that we all breathe the same air, are created from the same cosmic inspiration, and that our fates are intertwined.

The twin operating systems of democracy and capitalism are both designed to generate the greatest possible opportunity for all to experience life, liberty, prosperity, and happiness. When these systems operate from love, as they are meant to do, the result is inclusiveness, abundance, creativity, joy, and fulfillment for all. When these systems are dominated by fear, the result is oppression of minorities, attempts to suppress the right to vote, vilification of immigrants, greed, exploitation, and a growing disparity between winners and losers.

Business is poised to play the critical role in reorienting the magnet of the soul to the polestars of truth, beauty, and goodness. In the process it can drive the innovation that can heal many of our greatest social, economic, environmental, and political challenges. For this to happen, individual business leaders must experience an awakening of their consciences and choose to operate from love.

WHAT DO WE MEAN BY HEALING?

Webster's Dictionary defines *healing* as "to make free from injury or disease: to make sound or whole; to make well again: to restore to health; and to cause an undesirable condition to be overcome." The word *healing* shares its etymology with the words *whole* and *holy*. *Whole* means complete, unhurt, healthy, and undamaged. *Holy* means consecrated, sacred, godly, that which cannot be transgressed or violated.

Most human beings are wounded and their psyches are fragmented. They yearn to become more whole as they progress through life. What if the workplace could help fulfill that yearning? We will share stories of organizations that are doing just that: healing employees, customers, communities, and society while outperforming their peers financially.

We will show that it is usually not the work itself that is the cause of suffering; it is the *way* in which we work, the way we organize, manage, and lead. The exact same work can be a source of suffering and stress, or it can lead to fulfillment and flourishing—not only for employees, but for their families as well.

It doesn't matter what kind of work you do; you can have a healing impact. But, if you do not consciously choose to be part of the healing, you *will* be part of the hurting.

The first step in healing is bearing witness and being present to the reality that business often causes a great deal of unnecessary suffering. As James Baldwin said, "Not everything that is faced can be changed, but nothing can be changed until it is faced."[19] When we become aware of the ways in which we are causing suffering and empathize with those who are hurting, we can use our extraordinary creative capacity to find better ways to do business.

•   •   •

A traditional business says, "Here is an opportunity to make money by exploiting a need or gap in the marketplace."

A business with a slightly more advanced mindset says, "Here is an opportunity to make money by exploiting a need or gap in the marketplace, and we will initiate some corporate social responsibility initiatives and employee wellness programs to help mitigate the suffering we cause. And we will throw some money at a few charities."

A more evolved conscious business leader says, "Here's an opportunity to make a profit while serving customer needs and the needs of all stakeholders, including our communities and the environment."

A Healing Organization says, "Our quest is to alleviate suffering and elevate joy. We serve the needs of all stakeholders, including our employees, customers, communities, and the environment. We seek to continually improve the lives of all stakeholders while making a profit so that we can continue to grow and bring healing to more of the world."

Sometimes, businesses actually begin with this kind of noble healing purpose. For example, Mark Donohue, founder and CEO of

LifeGuides,[20] a Public Benefit Corporation, observed that a significant amount of suffering, even in a developed economy like the US, is related to what he terms "life challenges"—situations like caring for a mother with Alzheimer's, having a child with an addiction, going through a divorce, or having one's home burn down in the California wildfires. As he thought about the prevalence of these problems, he realized that there are many people who have dealt successfully with these difficult situations. He thought, "Perhaps there is a way to connect those in need with those who could help."

Shortly thereafter, at a business conference focusing on mission-driven technology entrepreneurship, he met the founder of Match.com and an idea emerged: What if he could create an enterprise that pairs people who are going through life challenges with trained "Peer Guides" who have successfully overcome the same difficulty? Just as Match.com pairs people for love, and Doctors On Demand finds physicians for patients, Mark visualized creating "a platform for caring people to do extraordinary good," and for those life guides to get paid a living wage for their efforts. As he states: "Life experience is the most valuable asset on the planet, yet we do a poor job of optimizing it. LifeGuides restores precious human connections, which have been fragmented by cultural and technological 'progress.'"[21]

In other cases, companies are transformed into Healing Organizations when the leader experiences an awakening of conscience. In part 2 of the book, we will share the stories of a wide range of healing leadership journeys. If you find yourself as moved by them as we did when we interviewed the leaders involved, it will forever change your life and your idea of what's possible.

In part 3, we will share a summary of what we have learned about how you can become a healing leader and help save the world through this transformational approach to business.

THE MORE BEAUTIFUL WORLD
OUR HEARTS KNOW IS POSSIBLE

Charles Eisenstein, author of *The More Beautiful World Our Hearts Know Is Possible,* talks about how people try to start change movements by creating a big wave. But he asks, "You can make a big wave but how do you change the deep currents?"[22]

In part 1, we will provide a broad historical perspective on the deep currents that are preventing us from moving toward a more human-centered approach to business. The deeper currents have to do with human nature itself. Are we wired to be purely self-serving creatures, or are we in fact wired to care? The answer is: both. In *Nonzero: The Logic of Human Destiny*, Robert Wright cautions,

> God knows greed won't vanish. Neither will hatred or chauvinism. Human nature is a stubborn thing. But it isn't beyond control. Even if our core impulses can't be banished, they can be tempered and redirected.[23]

Our core impulses also include love and compassion. There is no greater power or source of strength in the world than love. As His Holiness The Dalai Lama has said, "Love and compassion are necessities, not luxuries. Without them, humanity cannot survive."[24]

Never has the opportunity for positive impact been greater or more urgently required. If we fail to transform our approach to capitalism, then we will, as Charles Dickens prophesized:

> Crush humanity out of shape once more, under similar hammers, and it will twist itself into the same tortured forms. Sow the same seed of rapacious license and oppression over again, and it will surely yield the same fruit according to its kind.[25]

But, if we seize the moment and commit to live from our highest ideals and the core of our values, then Dickens's other prophecy will be realized:

> I see a beautiful city and a brilliant people rising from this abyss, and, in their struggles to be truly free, in their triumphs and defeats, through long years to come, I see the evil of this time and of the previous time of which this is the natural birth, gradually making expiation for itself and wearing out.[26]

The choice is ours.

THE
HEALING
ORGANIZATION

PART 1

BUSINESS— THE CAUSE AND SOLUTION OF MANY OF THE WORLD'S PROBLEMS?

RE-DREAMING
THE AMERICAN DREAM

Despite the proliferation of books on the decline of the United States and the advent of a post-American world, the US continues, more than any other country, to influence the course of world events, economics, and culture—in both positive and negative directions. It remains fair to say that "As goes America, so goes the world." And the "American Dream"—the idea that an ordinary person can achieve success through entrepreneurship, creativity, and hard work, rather than through inheritance, class, or caste—is now a universal dream, inspiring people in Bangalore and Beijing as much as it does in Bangor and Butte.

Modern democracy and capitalism took root in the United States, evolved here, and then spread to other parts of the world. Despite difficulties and setbacks, these two operating systems remain the twin hopes for human welfare.

But we are at an inflection point, a critical juncture in history where we must awaken conscience and consciousness to evolve these operating systems to meet the crises of our time.

Business is poised to play the key role in this evolution that can heal our planet and provide greater prosperity, abundance, health, and happiness for millions of people who are suffering needlessly.

Let's consider a historical perspective on how we got to this inflection point.

ENLIGHTENMENT WISDOM AND THE GREAT LAW OF PEACE

Enlightenment wisdom came out of the European intellectual movement in the seventeenth and eighteenth centuries, emphasizing reason and individualism over tradition and doctrine. The United States was founded as an expression of Enlightenment ideals such as freedom, individualism, reason, and egalitarianism, married with a deep sense of Judeo-Christian ethics. The English author G. K. Chesterton (1874–1936) called the US "a nation with the soul of a church."[1]

The story began with royal grants to fortune seekers as well as pilgrims coming from Europe to escape religious persecution. Over time, they were joined by huge influxes of economic migrants from all over the globe. The quest for freedom defined the ethos of what would become the United States. Initially about religious liberty, political and economic freedom soon became equally important. These became the country's defining qualities, which made it unique in the history of the world: a combination of enlightenment wisdom, pioneer spirit, and religious puritanism.

The "founding fathers" imbued a strong positive masculine energy into the Declaration of Independence, a beautiful and inspiring document by any standard.

The Declaration was about freedom and self-determination, but it was limited to white men. Although it was common usage at the time to refer to "Man" or "Mankind" as terms that referred to men and women, the founders didn't go out of their way to reference inclusiveness, or our collective responsibility to each other.

As the founders thought about how to create a system of government that could successfully deploy the ideals in the Declaration, they searched for historical precedents to guide them. They looked first to European models, some of which were hybrids of monarchies and varying degrees of democratic representation. They also looked at examples from Greco-Roman times, and even from the Bible. But everywhere they looked, they found war, conflict, and discord—until they looked right in their own backyard.

There they found a functional, democratic, decentralized governing system known as the Great Law of Peace. This had been put into effect by an alliance formed by five (later six) native American tribes in the Great Lakes region (present-day New York, northern Pennsylvania, and eastern Ontario). This Iroquois Confederacy was a sophisticated system that had worked well to reduce conflict and promote collective well-being since its establishment in the late sixteenth century between tribes that shared the Iroquois language.[2]

The Great Law of Peace protected freedom of speech, freedom of worship, and the rights of women. The tribes instituted three branches of government—including two houses and a grand council—and originated the notion of the separation of powers and checks and balances across different institutions. Notably, it had a Women's Council, analogous to the Supreme Court, charged with settling disputes and adjudicating legal violations by any of the member tribes.[3]

The founding fathers borrowed liberally from the Iroquois Confederacy for the government system they were designing for the fledgling country. At the Albany Congress in 1754, attended by representatives of all six tribes and seven of the colonies, Benjamin Franklin presented his "Plan of Union," in which he stated:

It would be a strange thing if six nations of ignorant savages should be capable of forming such a union, and yet it has subsisted for ages and appears indissolvable, and yet a like union should be impractical for ten or a dozen English colonies.[4]

At the time of the bicentennial of the US Constitution, the Senate passed a resolution stating, "The confederation of the original thirteen colonies into one republic was influenced by the political system developed by the Iroquois Confederacy, as were many of the democratic principles which were incorporated into the constitution itself."[5]

While historians debate the extent of the influence the Iroquois Confederacy had on the US Constitution, there were clearly two elements of the Great Law of Peace that the founding fathers did not embrace. The first was the "seventh generation" principle. Iroquois chiefs were required to consider the impact of every decision on the next seven generations. The second element they ignored was the empowerment of women. Though the leaders in the Iroquois Confederacy were men, known as *sachems*, they were always chosen by a council of clan mothers. Women owned the land and homes, and the Women's Council had veto power over any actions that could result in war. The council also had the power to impeach and remove any leader who was seen as acting improperly and appoint a new leader in his place.

Not only did the founding fathers not adopt any version of the Women's Council, they denied women the right to own property and to vote. The latter injustice would persist for another 140 years.

"REMEMBER THE LADIES . . ."

The US would have been well-served if it had founding mothers to go along with the founding fathers. The person who came closest to being a founding mother was Abigail Adams. In March 1776, three months before the Declaration of Independence was crafted, she wrote to her husband and future president, John Adams:

[I]n the new code of laws which I suppose it will be necessary for you to make, I desire you would remember the Ladies, and be more generous and favorable to them than your ancestors. Do not put such unlimited power into the hands of the Husbands. Remember all Men would be tyrants if they could. If particular care and attention is not paid to the Ladies we are determined to foment a Rebellion and will not hold ourselves bound by any Laws in which we have no voice or Representation. That your sex are naturally tyrannical is a truth so thoroughly established as to admit of no dispute; but such of you as wish to be happy willingly give up the harsh tide of master for the more tender and endearing one of friend.

While he respected Abigail and frequently sought her counsel, Adams considered this suggestion to be outlandish. In a response the following month (April 14, 1776), he dismissed it with a joke, flippantly suggesting that men feared the "despotism of the petticoat."

We have been told that our struggle has loosened the bonds of government everywhere; that children and apprentices were disobedient; that schools and colleges were grown turbulent; that Indians slighted their guardians, and negroes grew insolent to their masters. But your letter was the first intimation that another tribe, more numerous and powerful than all the rest, were grown discontented. Depend upon it, we know better than to repeal our masculine systems . . . . We have only the name of masters, and rather than give up this, which would completely subject us to the despotism of the petticoat, I hope General Washington and all our brave heroes would fight.

Abigail responded (May 7, 1776) with barely concealed fury:

Whilst you are proclaiming peace and good-will to men, emancipating all nations, you insist upon retaining an absolute power over wives. But you must remember that arbitrary power is like most other things which are very hard, very liable to be broken; and, notwithstanding all your wise laws and maxims, we have it in our power, not only to free ourselves, but to subdue our masters, and without violence, throw both your natural and legal authority at our feet.

The rebellion that Abigail Adams foresaw was a long time coming. Women finally got the right to vote in 1920, and the campaign for equal opportunity continues.

CAPITALISM HAD A FATHER *AND* A MOTHER

Many consequences flowed from those fateful early decisions, including the way capitalism evolved. As the US became the laboratory and model for representative democracy across the world, it also provided the most fertile soil for the spread of capitalism. This was rooted in the understanding that Adam Smith had given the world about the power of free markets. His great, abiding, and world-changing insight was that freedom leads to prosperity: Individuals pursuing their own perceived best self-interest and making decisions to further that would end up meeting each other's needs in a way that would be far more efficient and effective than if a central planning authority or governmental bureaucracy tried to make those decisions. The "invisible hand of the market" would allocate resources and effort and set prices through the ungoverned process of individuals making these little decisions. It would add up to collective well-being.

These ideas influenced the evolution of the American system directly. Benjamin Franklin dined with Adam Smith in Scotland in the early 1770s, well before *The Wealth of Nations* was published in March of 1776.[6] Smith read selections from the unfinished manuscript to Franklin, who would later bring some of that wisdom directly into the Declaration of Independence.

Smith's ideas became central to the defining identity of the United States: The government was not meant to occupy the "commanding heights" of the economy. It would be minimally involved in the market and people would be free to do what they wanted to do. The government's job was to create a level playing field (the rule of law) and get out of the way. This led to tremendous dynamism, more so than any other society had ever seen. The US was the first society in history where the "common man" (and it was limited to men) was in charge of his own destiny by right of law. Individuals could rise to great heights in their chosen field, through their own vision, creativity, daring, and hard work. You did not have to be born into the landed gentry or the aristocracy to have a chance at a good life. This country provided a greenhouse in which the conditions existed for humanity to make rapid progress in elevating living standards, and for individuals to be able to manifest and share their gifts.

The American experiment started with many healthy elements of masculine energy, embodied in strong, admirable qualities, such as love of freedom, self-determination, courage, resilience, self-expression, rugged individualism, achievement, and the drive for success. But, in the absence of countervailing feminine energy, this gradually and predictably became imbalanced and hyper-masculine. It manifested as domination, aggression, and excessive competitiveness that fueled a win-at-all-costs mentality.

The Great Law of Peace was predicated on a harmony of masculine and feminine energies and so was the original ideal of capitalism. Capitalism had a mother and a father; both were expressed in works of Adam Smith. Before *The Wealth of Nations*, Smith published *The Theory of Moral Sentiments* (1759) in which he proposed the ethical philosophy upon which capitalism and all societal institutions must rely. Smith understood capitalism as a system of human cooperation based on a balance of fundamental human motivations: self-interest and caring for others. We are not *just* self-interested creatures; that would render us sociopathic. We do things all the time that could be seen as going against our narrow self-interest because we care about others. Smith understood that human beings are driven by much more than self-enrichment and that capitalism needed a conscience.

MEN GONE WILD

After the US Civil War, industrial titans, or "robber barons" as they came to be called, emerged on the national stage to propel the spread of shipping and then railroads (Cornelius Vanderbilt), steel (Andrew Carnegie), oil (John D. Rockefeller), and banking (J. P. Morgan). The business sector grew massively and had most of the power in the US. Business leaders were called on to bail out the federal government from financial crises at various times, such as the Panic of 1893 and the Panic of 1907.[7]

Checkbooks and balance sheets were overwhelming the anemic system of checks and balances. There was no feminine moral sentiment to bring in nurturing, caring, or inclusive energy, and the government seemed powerless to regulate business in the collective interest of society. It was Men Gone Wild. Businesses became increasingly callous towards workers.

Carnegie Steel was a classic case. Carnegie's wealth, while enormous, had fallen behind that of Rockefeller in the early 1890s, largely due to a decline in the price of steel. Carnegie's right-hand man, Henry Frick, came up with a plan in 1892 to increase profit margins by significantly lowering labor costs. Frick was also looking to "break" the increasingly powerful Amalgamated Association of Iron and Steel Workers, which had been formed in 1876, and had been successful in negotiating better wages for steel mill workers. Their contract at the Homestead plant was up, and Frick announced that he would no longer negotiate with the union, only with individual workers. He issued an order: Workers would now be required to work six days a week instead of five, and twelve hours a day instead of ten. And, their pay would be reduced.

To the workers, this was adding insult to real injury. According to Carnegie's biographer Peter Krass, 20 percent of all male deaths in Pittsburgh in the 1880s were due to fatal accidents in the steel mills.[8] Approximately 9 percent of steelworkers were literally dying on the job each year, many collapsing from exhaustion.[9]

When the workers protested, Frick called in the Pinkerton National Detective Agency, which around that time had more men and guns than the US Army. When the Homestead workers refused to back down, the Pinkertons opened fire: Nine workers were killed and forty wounded. That led to retaliatory violence; seven Pinkerton detectives were killed and twenty wounded.[10] An anarchist unconnected to the union later stabbed Frick and nearly killed him. Eventually, Carnegie prevailed, as the union was broken, wages were lowered, working hours increased, and profitability rose. Appalled by the violence on both sides at Homestead, twenty-six states passed laws against hiring outside guards.

Carnegie had once championed the rights of workers and supported the idea of unions and his conscience was reawakened by these unfortunate events. In his autobiography, he wrote, "Nothing . . . in all my life, before or since, wounded me so deeply. No pangs remain of any wound received in my business career save that of Homestead."[11]

This ugly episode contributed to a hardening of the divide between labor and management. To many people, that divide seems natural; labor and management have come to be seen as natural enemies. But why should the employees of a company be enemies with its founders and leaders? The divide is based on a pseudo-Darwinian idea of competition for a limited amount of resources; if management has more, labor has to have less and if labor has more, there will be less profit. It is a fear-driven, scarcity-based "zero-sum" way of thinking about business. It views owners and workers as combatants over a limited pool of financial wealth, rather than as co-participants in a system of multifaceted value creation.

The growing gulf between labor and management created fertile ground for Marxism to spread, followed by socialism and communism, which in turn led to an extraordinary amount of strife and suffering all over the world for the next hundred years.

We believe that a caring rather than warlike approach to commerce could have avoided this whole epoch in our history.

It is important to acknowledge that Vanderbilt, Carnegie, Rockefeller, and Morgan were exceptional human beings. A lot of what they accomplished was an expression of the healthy masculine energy. None of them came from privileged backgrounds. They had the audacity to believe they could become extraordinarily wealthy while transforming society through their expansive and bold visions. But they all got carried away by a compulsive drive to dominate and maximize their own wealth at any cost.

THREE PERSPECTIVES ON THE "BUSINESS OF BUSINESS"

Like Carnegie and Rockefeller, many well-meaning people still inhabit an arid "zero-sum mindset" world, believing that business is mostly about competition and that resources, including other humans, are there to be exploited. Then, as they approach the end of their life, they become charitable and give much of their fortune away to make up for the harm they've caused. This way of thinking leads to great suffering for all parties, including the earth itself.

There is a better way. You can do good all along the journey, enrich people's lives, and help save the planet—*while making more money*. Once you realize that's possible—and we will show you that it is happening now in a range of industries around the U.S. and the world—we believe you'll want to be part of and do business with companies with this awakened awareness. For this to happen, we need to transcend the outdated divides between management and labor, between conservatives and liberals, and between those who continue to promote unfettered capitalism and those who believe that socialism is the only answer to corporate greed and corruption.

A few years before the Great Depression, President Calvin Coolidge gave a speech in which he said, "The chief business of the American people is business."[12] His remark captured something true and positive; the US came into being as a uniquely entrepreneurial nation. American kids never grew up dreaming of civil service jobs or inheriting titles. But Coolidge's remark begs the question, "What is the business of business?" This is where, for close to 150 years, we got trapped in a narrow, selfish, and instrumental attitude, encapsulated in Alfred P. Sloan's statement that "the business of business is business."[13]

Sloan built General Motors into the world's largest corporation, and invented many aspects of modern management. He believed that business existed purely to make money, and anything else was a distraction at best. This viewpoint was buttressed by economist Milton Friedman when he wrote his momentous 1970 essay in the *New York Times Magazine* titled "The Social Responsibility of Business Is to Increase Its Profits." This perspective became dogma. Most business school curricula are built on that foundational premise. But it is a deeply flawed premise, and has caused massive, unnecessary suffering to people and the planet.

There is a third perspective on the nature of the business of business. It comes from Herb Kelleher, the leader who built Southwest Airlines into the world's most successful airline, and into one of the most admirable businesses of any kind. Founded in 1967 to offer service between cities in Texas, Southwest went public in 1971, under the stock market symbol LUV. The company manifests love, care, and joy in a multitude of ways, and it has been consistently profitable every year since it started operations. It has never had a strike, despite being heavily unionized. Southwest encourages employees to "Live the Southwest Way." This includes "having a Warrior Spirit (work hard, desire to be the best, be courageous), a Servant's Heart (put others first, live by The Golden Rule), and a Fun-LUVing Attitude (be a passionate teamplayer, have fun)."[14]

Kelleher, who passed away while we were writing this, coined the phrase that represents the core of the philosophy of the Healing Organization: "The business of business is people—yesterday, today, and forever."

• • •

CREATING A NEW MYTH FOR BUSINESS

The dominant narrative about business remains focused narrowly on generating as much profit and growth as possible. This is done by generating as much revenue as possible, which means selling as many products as possible to as many people as possible at as high a price as you can get away with, whether people benefit from these products or not. Since profit equals revenue minus cost, traditional businesses also look to minimize costs. They do that by spending as little as possible on their people, squeezing their suppliers to the extent that they can, externalizing any costs that they can onto society and the environment, and minimizing the taxes that they pay.

It's time to create a new, more human, life-affirming story about what business is and what it can be, one that can serve us and will heal us. In *The World Is As You Dream It*, economist John Perkins recounts a conversation he had with a Numi, a shaman in South America:

JP: I sometimes think that all we care about is money and dominating things. Other people. Other countries. Nature. That we have lost the ability to love.

NUMI: You haven't lost the ability. . . . The world is as you dream it. Your people dreamed of huge factories, tall buildings, as many cars as there are raindrops in this river. Now you begin to see that your dream is a nightmare.

JP: How can my people change this terrible situation we've created?

NUMI: That's simple. All you have to do is change the dream. It can be accomplished in a generation. You need only plant a different seed, teach your children to dream new dreams.[15]

The Healing Organization is a new dream about business.

What if instead of causing or exacerbating many of our greatest problems, business could be the solution?

What if we integrated and applied the wisdom of *The Theory of Moral Sentiments* in concert with *The Wealth of Nations*?

What if business could be the path to healing our society and restoring the American Dream—not only for Americans but for all of humanity?

We believe it can and it must. And for this to happen, we need a new narrative based on an integration of our archetypal energies.

INTEGRATING THE FOUR ARCHETYPAL ENERGIES

Human families have four archetypal roles: father, mother, child, and grandparent. Correspondingly, human culture is motivated and sustained by four fundamental energies: masculine, feminine, child, and elder. When healthy, these energies manifest as achievement, caring, joy, and purpose. Each represents a crucial and complementary piece of what we need to live healthy, meaningful, and joyful lives.

Many human societies all over the world were originally matriarchal, and attuned to and reverent of the earth. But over time, because of plagues, disease, and famines, a premium was put on the ability to control the environment. Evolutionarily, the aggressive, analytical, bolder elements began to dominate.

Traditional societies also revered their elders and relied on the wisdom of those elders. But beginning with the Industrial Revolution, fast-evolving industries and markets demanded families be more mobile to take advantage of job opportunities. Elders were often left behind. In 1924, Alfred Sloan introduced the strategy of *planned obsolescence* that began with building cars that would break down right around the time the warranty expired and then aggressively marketing new models. This helped foment a mania that continues into the present for things that are "new" and "the latest model," resulting in an attitude of planned obsolescence toward humans, who are often forced to retire from the workforce based purely on age.

Martin Luther King Jr. spoke of the need to be "tough-minded and tenderhearted"—a blend of the healthy masculine and feminine energies.[16] We also need to be wise: connected to our principled wisdom, our higher selves, and thus steeped in elder energy. We need to be playful, to retain our capacity for joy, fun, and creativity. This means we have to be in touch with our joyful, healthy child energy. All of these energies are alive and harmonized in a Healing Organization.

Herb Kelleher brought these energies in harmonious proportion to the creation of the culture at Southwest Airlines. Herb embodied and cultivated the elder energy by defining and sharing a higher purpose for his enterprise: to democratize air travel. His healthy masculine energy is expressed in Southwest's Warrior Spirit. Trained as an attorney, Herb won legendary legal battles against firms who tried to destroy his innovative enterprise, and he ran the business with remarkable acumen and innovative thinking. His healthy feminine energy is beautifully alive in the company's logo (a heart with wings) and in the caring and kindness (the Servant's Heart) with which Southwest employees treat one another and their customers. And, delightfully, Kelleher's healthy child energy lives in the FunLuving culture where flight attendants have been known to pop out of overhead compartments to lead passengers in song and where safety announcements sometimes sound more like rap songs and stand-up comedy routines.

The opposite is a toxic leader who simultaneously manifests the unhealthy expression of each of these four energies. More common are fragmented leaders who are only able to access one or two of these energies, and not always in their healthy expressions. The same thing applies to organizations, as well as countries. Thriving organizations manifest the four energies in their healthy expressions, while dysfunctional organizations do the opposite.

AN AMERICAN NIGHTMARE?

The United States began its existence with positive masculine energy (rooted in freedom and self-determination) and wise elder energy (reflected in religious principles guiding personal and work conduct). But without the balancing aspects of caring and playfulness, these devolved into domination and dogma. In recent generations we've witnessed an upsurge of unhealthy child/adolescent energy manifest in the readiness that many have to be offended by anything that "makes them uncomfortable" and the unrestrained compulsion toward immediate gratification as expressed in ever climbing levels of personal and governmental debt.

The result is that, for many, the American Dream has become a nightmare.

America leads the world in obesity, gunning down schoolchildren, opioid addiction, and percent of the population in prison. Although individuals must bear responsibility for eating junk food, shooting their neighbors, snorting OxyContin, and committing crimes, all of these national crises are enabled and exacerbated by business. Food scientists are engaged to craft unhealthy items that pander to and exploit deep-seated cravings; firearms trade associations fight against any limitations to the ability of anyone, even the mentally ill, to purchase and deploy a deadly weapon; pharmaceutical companies promote addictive drugs using strategies similar to those used by illegal cartels; and for-profit prisons greedily welcome new "customers."

The most heartbreaking statistic is that the suicide rate in the United States increased by 33 percent between 1999 and 2017, even as it declined 29 percent globally between 2000 and 2018.[17]

In *The Unwinding: An Inner History of the New America*, George Packer paints a vivid picture of this nightmare:

> He was seeing beyond the surfaces of the land to its hidden truths. Some nights he sat up late on his front porch with a glass

of Jack and listened to the trucks heading south on 220, carrying crates of live chickens to the slaughterhouses—always under cover of darkness, like a vast and shameful trafficking— chickens pumped full of hormones that left them too big to walk. He thought how these same chickens might return from their destination as pieces of meat to the floodlit Bojangles' up the hill from his house, and that meat would be drowned in the bubbling fryers by employees whose hatred of the job would leak into the cooked food, and that food would be served up and eaten by customers who would grow obese and end up in the hospital in Greensboro with diabetes or heart failure, a burden to the public, and later Dean would see them riding around the Mayodan Wal-Mart in electric carts because they were too heavy to walk the aisles of a Supercenter, just like hormone-fed chickens.[18]

Behind each element of Packer's vignette there was a conscious business decision that somebody made with an eye toward maximizing profits without regard to the welfare of employees, "consumers," or society.

All of this madness has to stop, but relying primarily on government regulation just doesn't work; it tends to punish all businesses, even those who are genuinely serving real human needs in a thoughtful and caring way.

BUSINESS CAN HELP SAVE THE WORLD AND RESTORE THE AMERICAN DREAM

Jonathan Haidt, professor of ethical leadership at NYU, makes the point that capitalism promises, and delivers, on dynamism, while socialism promises, but is not able to deliver, on decency. Since we know that socialism doesn't work and runs counter to our desire for freedom and self-determination, many people believe that we are therefore consigned to live an existence of dynamism *without* decency: a world of ruthless competition, in which human beings are valued only to the extent that they are able to generate more output for less than the person next to them.

In the big picture of human history, thus far, this form of unconscious capitalism has, nevertheless, done more good than harm.[19] But it also causes massive unnecessary suffering. And it is no longer sustainable. The natural resources necessary to fuel it are finite and dwindling and the effects of their over-exploitation threaten to fry, asphyxiate, starve, and drown us.

Our current model that puts shareholder interests above all other concerns has resulted in the sacrifice of decency for dynamism. People who want to "do good" look for careers in nonprofit organizations while most just accept that business without a higher purpose, business that exploits people and the planet, is, well, business as usual. The notion that goodness and profit aren't compatible is a false and tragic assumption predicated on a psyche disordered by imbalanced energies and an incomplete reading of Charles Darwin and Adam Smith.

Choosing between caring for others and self-interest is like choosing between breathing in and breathing out. Think of inhaling as taking in something from the world. Think of exhaling as giving back something to the world. We breathe in oxygen, which trees breathe out. We breathe out carbon dioxide, which the trees breathe in. We can have a world in which we can all breathe fully and freely, with a harmony of dynamism and decency. That is not a utopian ideal. It's a reality in the businesses you will learn about in this book. These businesses have *greater dynamism* than traditional profit-centered companies. They offer true decency: not the indignity of receiving a handout from the government or the indecency of a dehumanized work environment, but the beauty and satisfaction of a life lived with purpose, caring, abundance, and joy.

• • •

Before the year 2050, we will have to reinvent the ways in which we meet virtually all of our needs. That calls for an extraordinary amount of innovation, which we know from history only the dynamism of a free market system can deliver. Fortunately, our capacity for innovation is vast and can be developed further. As much as it is inspired by the desire for individual gratification, it is motivated much more profoundly when we are focused on a higher purpose in service to our fellow humans.

When we focus our businesses on meeting genuine human needs, creating value, and being of service to one another, we can fulfill our own self-interest in a much richer, deeper way. It's time for capitalism to evolve beyond the false trade-offs we have created between caring and succeeding, between love and profits.

Previous generations didn't know that this was possible. Now, it is clear: Companies that operate from a higher purpose beyond just making money will thrive. Companies that treat all their stakeholders with care will thrive. These "Firms of Endearment" are not only much healthier places to work, they are loved by their customers and communities and they are *more profitable* than companies that focus solely on financial return.[20]

Once you know that this is possible, why would you consider doing anything else? How could you settle for anything less?

Of course, it's more than just a matter of reading research and understanding the business case for a more caring and balanced approach to commerce. It's about aligning your own purpose with the higher goal of making the American Dream real again for everyone.

The Healing Organization is a book of stories about companies aspiring to live this evolved dream of what business can be. Although none are perfect, they operate in a way that generates engagement and fulfillment for their employees, delight and loyalty for their customers, positive contributions to their communities and to the environment, and excellent returns for their owners and investors.

In order to embrace this evolving model, it helps to have an even greater historical perspective on how organizations evolved and on how they can bring out the worst or the best in human nature.

EVOLVING FROM EMPIRE TO MINISTRY, FROM CONQUERING TO CARING

W hat's the greatest empire in human history? It wasn't the Holy Roman Empire, which as Enlightenment philosopher Voltaire (1694–1778) quipped, "Wasn't holy, wasn't Roman, and wasn't really an empire." Contenders include the Egyptian, Roman, Mongol, British, Spanish, Mughal, Russian, Persian, Qing, and Ottoman empires. Historians argue about the criteria for "greatness" debating about the weighting factors: amount of territory under domination (British), percentage of the world's population under its control (British), and duration (either the Ottomans or the Romans depending on whether you count the Eastern Empire as truly Roman).

All of these empires have influenced humanity culturally and many of those influences benefit us today, but all of them were established and maintained through violence and the imposition of slavery on conquered peoples. Emperors, shahs, caesars, kaisers, tsars, pharaohs, and kings all believed that they were born with a mandate to conquer others and expand their domination, whatever the cost in human suffering.

The result is that since the first Egyptian empire was established approximately 4,700 years ago, humans have been at war most of the time. Estimates by historians vary but the general consensus appears to be that the world has been without a significant war for less than 10 percent of recorded history. European nations fought 1,200 wars with each other in the 600 years leading up to the end of World War II.[1] In the twentieth century alone, more than 120 million people were killed in the attempt to expand or defend territory.

In assigning responsibility for all the carnage and suffering, we can't underestimate the role of individual leaders, driven by egotism, greed, and the lust for power and glory, and supported by their tribe's belief that their god or ideology is superior to all others. Kings and emperors aspired to earn the historically dubious distinction of being labeled "Great." All that usually means is that they caused great suffering, in vain attempts to fill a void in their psyches that can never be filled by any amount of power or wealth. Many ascended to their thrones by killing siblings, some by killing parents. If successful, by their own perverted definition of success, they lived to see their empires span vast geographies, subjugating other tribes as well as their own people. They built ostentatious towers, walls, and palaces in which they spent their last years basking in the fear-induced subservience of those around them. And then they died, or more likely, were murdered by another bloodthirsty, power-hungry would-be monarch.

The amount of suffering caused by power-crazed empire builders on ordinary, innocent human beings is almost unimaginable.

"GREAT" IS RARELY GOOD: THE LANGUAGE OF EMPIRE

And yet, for many businesses, the ethos and language of empire, conquest, and domination continues to influence how they think about what they do and why they do it. We strive to "capture market share" and celebrate "making a killing." We refer to competition as cut-throat, and people who are supposed to be leaders as bosses, a term with roots in slavery and the Mafia.

One of the "sharks" on the popular *Shark Tank* television show, who is held up as a role model for would-be entrepreneurs, counsels: "Business is war. . . . I want to kill the competitors. I want to make their lives miserable. I want to steal their market share. I want them to fear me."[2] He adds, "Working twenty-four hours a day isn't enough anymore. You have to be willing to sacrifice everything to be success-ful. . . . You may lose your wife, you may lose your dog, your mother may hate you. None of those things matter. What matters is that you achieve success and become free."[3]

Does this sound to you like a healthy, sustainable idea of success and freedom? Is this the world you want your children to grow up in?

The business and leadership sections of bookstores everywhere usually feature titles that glorify this sociopathic approach to business as empire and CEO as emperor. In an article about the ten worst business books of all time, *Inc.* magazine contributing editor Geoffrey James chose *Leadership Secrets of Attila the Hun* as the absolute worst. He writes, "CEOs are supposed to provide a service to their firms, not emulate dictator thugs."

This genre also includes titles like:

- *Alexander the Great's Art of Strategy: The Timeless Lessons of History's Greatest Empire Builder*
- *Napoleon on Victory, Leadership and the Art of War*
- *How to Become an Apex Predator Level Billionaire*
- *Way of the Wolf* (Yes, written by the actual "Wolf of Wall Street")
- *I'll Make You an Offer You Can't Refuse: Insider Business Tips from a Former Mob Boss*

Subtler works by Sun Tzu and others offer valuable guides to strategy—the ancient Chinese master strategist counsels, "It is best to win without fighting," and, "A leader leads by example not by force." But too many businesses have developed an overreliance on metaphors derived from empire building, war, predation, and organized crime.

"GOOD" TO "GREAT" ISN'T EITHER . . .

In his bestseller *Good to Great: Why Some Companies Make the Leap . . . and Others Don't,* Jim Collins profiled companies that, according to his criteria, would serve as ideal role models for business.[4] From the previous hundred years of publicly traded companies, only eleven made the list. Collins defined "greatness" along a single dimension: financial performance. He highlighted publicly traded companies that were "good" because they were performing financially at an average or good level, and then became great, i.e., they performed financially at a higher level (they outperformed the market by at least 3 to 1 over at least a fourteen-year period).

The question is, "How 'good' were they to begin with?"

The list included Circuit City, Fannie Mae, Wells Fargo, and Phillip Morris.

Before it went bankrupt, Circuit City dismissed all of its hourly employees who were earning above minimum wage so they could replace them with people making minimum wage, regardless of performance.

Fannie Mae was a significant perpetrator in the financial crisis of 2008–09, and Wells Fargo keeps racking up massive fines for unethical business practices.

Philip Morris was the world's largest tobacco company at the time the book debuted in 2001. Life expectancy declines by thirteen years for heavy smokers, seven million people die every year from the effects of tobacco, and we spent about $422 billion in 2012 to deal with the public health consequences—that's the industry's impact on the world.[5] The better Philip Morris gets at its job, the more humanity suffers. If financial performance is your only metric, chances are that you are just causing more suffering for more people more efficiently.

It's not great. It's not good. It's insane.

IT'S ALWAYS PERSONAL: A NEW WAY OF DOING BUSINESS

The Godfather movie turned the line "It's not personal, it's just business" into a meme, later adopted by the television show *The Apprentice*. Originated by mob accountant Otto "Abbadabba" Berman (1891–1935), who was gunned down at the Palace Chophouse restaurant in Newark, New Jersey, in 1935, the phrase is commonly used to justify self-aggrandizing, morally bankrupt, and inhumane behavior.[6] Yes, leaders must sometimes make tough decisions that may negatively impact some stakeholders, but genuine leaders do so with awareness that their actions are profoundly personal to those affected. They act with compassion and kindness.

Given the human history of brutality and murder, it is worth considering how we can change our metaphors, our language, and our consciousness about business so that we can do a better job of creating wealth while also promoting well-being. This can't be done naively. A healing leader must cultivate the balance of compassion and kindness with toughness and business acumen. Healing leaders must first know, in the words of another classic title, how to *Swim with the Sharks without Being Eaten Alive* while going beyond the consciousness of the shark tank.

Paul Polman, who retired as CEO of Unilever at the end of 2018, is an inspiring example. Under his decade-long stewardship, Unilever began to transform from an empire into a Healing Organization, with a stated purpose of "making sustainable living commonplace." Polman helped craft this healing aspiration for the company:

Our vision is a new way of doing business—one that delivers growth by serving society and the planet.[7]

He states,

We have to bring this world back to sanity and put the greater good ahead of self-interest. We need to fight very hard to create an environment that is more long term focused and move away from short termism.[8]

In 2017, Unilever was forced to fight for its life when it was targeted for acquisition by 3G Capital, an investment firm based in Brazil that some regard as a contemporary embodiment of the scorched-earth philosophy of a Mongolian khan. Starting in 1997, 3G has engineered nearly $500 billion of deals, creating multiple "platforms" through which it has consolidated businesses and created global behemoths in beer (AB-InBev), processed foods (Kraft Heinz), and fast-food restaurants (Restaurant Brands International, which includes Burger King and Tim Hortons). The company operates with a single-minded devotion to managing by the numbers. Dedicated to producing the absolutely highest profit margins in the shortest time, the company eliminates as many jobs as possible and then focuses relentlessly on cutting every cost every year.

Managers who embrace this cutthroat culture stand to make a lot of money—as long as they keep delivering the numbers. They know they will be replaced the instant they slip. "We're constantly trying to train new people and we're constantly telling everybody that the newer people should be better than the old people," says 3G lead partner J. P. Lemann.[9]

What is a "better" person? For 3G, it is somebody who is even more ruthless about cutting costs and increasing margins, without regard for human consequences or for the future.

If you feel exhausted and demoralized simply reading about this culture, you're not alone. The good news is that the moral bankruptcy of this approach to business is starting to be exposed. Yes, profit margins can be increased in the short term through massive layoffs, choking off investments designed to yield returns in the future, and aggressively marketing what you have today, but eventually this approach is unsustainable. Every empire eventually faces its day of reckoning. After treating people as soulless, disposable automatons, 3G's "dead companies walking," like termite-infested houses, appear to have been hollowed out. Revenue growth has stalled and by March 2019 the share price for Kraft Heinz had declined 66 percent from its peak in 2017, while Ab-InBev fell 38 percent. In a recent interview with a *Fortune* executive editor, the seventy-nine-year old Lemann, whose personal net worth peaked at $30 billion (it is $22.4 billion at this writing), described himself as "a terrified dinosaur."[10]

Paul Polman, who is not given to hyperbole, describes the campaign to fend off the Tyrannosaurusian attempt to eviscerate Unilever as a "near-death experience." Polman's approach to creating long-term value for all stakeholders is expressed in these remarks:

> I don't think our fiduciary duty is to put shareholders first. I say the opposite. What we firmly believe is that if we focus our company on improving the lives of the world's citizens and come up with genuine sustainable solutions, we are more in synch with consumers and society and ultimately this will result in good shareholder returns.[11]

During his ten-year tenure at the helm of this century-old company, Paul Polman didn't merely espouse high-minded ideals. Under his leadership, Unilever shares rose 150 percent, far in excess of the 70 percent rise in the benchmark FTSE 100 index. Unilever significantly outperformed its competitors such as Nestle and Procter & Gamble.[12]

It is the exact opposite of the strategy of 3G and others like them who focus exclusively on short-term return for themselves and their shareholders. Fortunately, with the support of his board of directors, Polman and Unilever thwarted the takeover attempt and reaffirmed their commitment to their goal of Transformational Change at the

System Level, which includes initiatives to heal the environment, feed the hungry, empower women, and eradicate poverty globally.

WHAT IF YOUR BOSS ACTUALLY IS A PSYCHO?

Who has orange hair, decimated multiple companies, wrote a book bragging about his prowess as a dealmaker, surrounds himself with gold, praises predators, and relishes firing people? Sadly, there are a number of bosses who fit the description above, but we are referring to "Chainsaw" Al Dunlap, a perennial fixture on any top ten list of the worst CEOs ever.

In a 2015 story entitled "Your Boss Actually Is a Psycho," *GQ*'s Jon Ronson quotes Dunlap: "Predators. Predators. I have a great belief in and a great respect for predators."[13]

Fast Company magazine editor John Byrne, author of *Chainsaw: The Notorious Career of Al Dunlap in the Era of Profit-at-Any-Price*, wrote:

> In all my years of reporting, I had never come across an executive as manipulative, ruthless, and destructive as Al Dunlap. . . . Dunlap sucked the very life and soul out of companies and people. He stole dignity, purpose, and sense out of organizations and replaced those ideals with fear and intimidation.[14]

Bad bosses seem to be present at many levels of business and elsewhere. Toxic systems foster toxic leadership, and toxic leaders bring out the worst in people. As John Byrne put it,

> And for every destructive and crooked boss heading off to jail, there are thousands of junior Captain Queegs out there. They poison the work environment, spreading fear and apprehension. They ridicule colleagues, have a grandiose sense of self-worth, and fail to accept responsibility for their own actions. Most of

them get away with it because they deliver the results—if only in the short term.[15]

For example, under the stewardship of John Stumpf, former CEO of Wells Fargo, employees opened more than 500,000 credit card accounts and 1.5 million checking accounts in the names of their customers, without the customers' permission, in order to generate phony sales commissions. The board fired Stumpf but replaced him with Tim Sloan, who was his deputy in presiding over the debacle. Under Sloan's watch, things further deteriorated. After paying $50 million to settle a lawsuit alleging that the bank gouged homeowners who were in default on their mortgages, Wells Fargo was fined $1 billion by The Consumer Financial Protection Bureau as punishment for gross irregularities in its mortgage and auto loan businesses. The agency filing noted: Wells Fargo's "conduct caused and was likely to cause substantial injury to consumers."[16] Sloan resigned abruptly in March of 2019.

One might ask: Are Dunlap, Lemann, Stumpf, Sloan, and their legions of "Junior Captain Queegs" bad people?

Our role isn't to pass judgment, but rather to question:

Could there be something wrong with the system in which they all operate? Something that brings out the worst in all of them, that exacerbates whatever moral blind spots they may have?

BAD PEOPLE OR BAD IDEAS?

Author and neuroscientist Sam Harris makes a crucial distinction between bad ideas and bad people.

I have a very strong sense, and there is a lot of evidence to back this up, that (most of) the mayhem we see created by people in the world is not the result of bad people doing bad things they would do anyway because they're bad. It is the result of good people—or at least psychologically normal people—under the sway of bad ideas. *Bad ideas are far more powerful than bad people.*[17]

Nobel Laureate Muhammad Yunus adds:

It is not because bad people are running the machine; just that the machine is built that way. The system was not designed to have any moral responsibility. This machine turns people into money-centric robots. Business schools compete to produce market-warriors.[18]

The realms of business and corporate governance have been under the sway of many bad ideas for a long time, or at least ideas that are now obsolete. A short list of these bad ideas includes:

- Human beings are driven only by self-interest and are purely rational economic value maximizers.
- Business exists only to maximize profits for its owners.
- Work only matters in people's lives to the extent that it generates income.
- The best way to motivate people is to use a combination of carrots and sticks.
- The job of the leader is to motivate, pressure, or coerce people into behaving in ways that achieve the leader's objectives.
- The world of work is separate and distinct from the world of our personal lives.
- The best way to increase profits is to squeeze employees and suppliers.
- It's acceptable to mistreat people and foul the environment if you donate lots of money to charity.

These ideas have been crystallized into theories, and have become dogma for many business leaders. But when money-centric robots aren't programmed with moral responsibility, the result is often evil.

THE LUCIFER EFFECT

On April 11, 1961, Nazi functionary Adolf Eichmann was put on trial for his role in systematically murdering millions of innocent civilians during World War II. Testifying from inside a glass booth to protect him from assassination, Eichmann claimed, unapologetically, that he was "merely a little cog in the machinery" of the death camps and that he wasn't responsible because he "was only following orders." On December 12, 1961, he was found guilty of mass murder and other heinous crimes and was sentenced to death shortly thereafter.

While the trial was going on, Stanley Milgram, a newly appointed professor at Yale University's Department of Psychology, began a series of experiments to understand how Eichmann, and millions of his countrymen, were able to participate in mass homicide. Were Germans more susceptible to authoritarian rule than others? "Could it be," he wrote, "that Eichmann and his million accomplices in the Holocaust were just following orders? Could we call them all accomplices?"[19]

Instead of trying this in Germany, Milgram recruited subjects locally in New Haven, Connecticut, for what they were led to believe was a study focusing on memory and learning. Clad in a white coat, the experimenter instructed the subjects to administer increasingly painful electric shocks to a learner when the learner made mistakes in a memory task. The experimenter (authority figure) and the learner were actors employed by Milgram. Subjects held the control lever for a shock generator with a gauge that began with 15 volts (labeled "Slight shock") increasing in increments to 375 ("Danger: Severe Shock") going all the way to 450 volts ("XXX"). They were instructed to intensify the shock with each error.

As the shock level went up, the learner, who could be heard but not seen by the subject, writhed in apparent agony and screamed in what seemed like terrible pain. As learners pleaded to be released and begged for the experiment to stop, subjects would often turn to the experimenter and express discomfort with the process and question whether they should continue. They were instructed to continue with the experiment, and were assured that if something happened to the learner, they would not be held responsible.

Before publishing his results, Milgram polled forty American psychiatrists asking what percentage of people they thought would go all the way to 450 volts. The consensus was that just 1 percent, the proportion of sadists in the general population, would deliver the fatal dose.

Over the course of a number of versions of the experiment, Milgram found that *up to two-thirds* of the subjects delivered what they believed to be lethal shocks to the learner, as they were instructed to do by the authority figure.

If you go online, you can see footage of the original Milgram experiments where seemingly normal Americans, who were paid just four dollars per day to participate, yield to authority, and torture their fellow citizens because someone in a white coat told them to do so and absolved them of responsibility for the consequences.[20]

Ten years later, Philip Zimbardo, a professor of psychology at Stanford University and Stanley Milgram's high school friend, conducted another experiment to answer the question framed in the trailer for the movie made about it:

> What happens when you put good people in an evil place? Does humanity win over evil, or does evil triumph?

Zimbardo began his study with what he describes as "an average group of healthy, intelligent, middle-class males." Earning fifteen dollars per day to participate, the young men were assigned, by the flip of a coin, to play the role of either prison guard or prisoner. Zimbardo emphasizes that role assignments were purely arbitrary, and that there were no differences between the subjects assigned to be guards and those assigned to be prisoners. The result?

> Our planned two-week investigation into the psychology of prison life had to be ended after only six days because of what the situation was doing to the college students who participated.

In only a few days, our guards became sadistic and our prisoners became depressed and showed signs of extreme stress.[21]

In his book *The Lucifer Effect: Understanding How Good People Turn Evil,* he explores the implication of this and other research into our capacity to be kind or cruel, caring or indifferent, creative or destructive—and what makes us villains or heroes.

When the atrocities at Abu Ghraib prison in Iraq were made public in April 2004, Americans were horrified that young men and women serving in the military, who we considered heroes, could participate in the gross abuse and torture of prisoners, especially as it seemed to be done for the amusement of the abusers, rather than any pretense of intelligence-gathering. Zimbardo explains that the dynamics that drove the revolting behavior at Abu Ghraib were the same as those in his prison experiment years before: Perceived anonymity, the absence of a sense of personal responsibility, and tacit approval or encouragement by authority figures are all elements that are usually present when seemingly good people do evil.[22] The same elements are at play when salespeople at Wells Fargo open fake accounts. People enter what Milgram calls an "agentic state," blindly following orders without regard to their consequences or moral implications.[23]

When something bad happens, the tendency is to blame a few bad apples, not the system. Zimbardo notes that most psychotherapists focus on individual psychopathology as the driver of bad behavior. This is the *Bad Apple theory*. Social psychologists look for situational explanations, focusing on the external forces that seem to encourage or dissuade bad behavior. This is the *Bad Barrel hypothesis*. Systems theorists seek broader explanations focusing on the cultural, political, economic, and legal structures that make people behave the way they do. This is the *Bad Barrel Makers hypothesis*.

Ultimately, good and evil actions arise through a complex dynamic interplay of all these elements. What do people bring into the situation? What does the situation bring out in them? What is the system that creates and maintains that situation?

As a classic *New Yorker* cartoon expresses it:

I'm neither a good cop nor a bad cop, Jerome. Like yourself, I am a complex amalgam of positive and negative personality traits that emerge or not, depending on circumstances.[24]

How can we leverage our understanding of human behavior to create organizational circumstances that bring out the best in the "complex amalgam" of humanity and avoid the worst?

Among the many powerful lessons from the work of Milgram and Zimbardo about avoiding a hurting culture and creating a healing one, two stand out:

- All evil begins with 15 volts! Small actions matter, for better or for worse.
- Leaders must set a positive example: In the Milgram experiments, 90 percent of subjects who witnessed another subject go all the way to 450 volts did the same, but when they saw someone rebel and refuse to administer more shocks, 90 percent also rebelled.

In other words, leadership really matters, and even the small actions of people in power have profound effects for better or worse.

THE ANTIDOTE TO EVIL: THE PATH OF THE EVERYDAY HERO

While we work on a better way of making barrels, we can all embrace everyday heroism as the antidote to evil. Zimbardo defines heroism as "intentional action to protect others without expectation of personal gain and with awareness of likely personal costs." He adds, "Heroes defend, uphold, and promote causes that benefit the greater good, despite pressures to do otherwise or potential risks to doing so."[25]

His Heroic Imagination Project is a global initiative with a mission: "To encourage and empower individuals to take heroic action during crucial moments in their lives. We prepare them to act with integrity, compassion, and moral courage, heightened by an understanding of the power of situational forces."[26]

Heroes are ordinary people who go out of their way to help others in need. The Healing Organization emphasizes the importance of individual moral responsibility while creating systems and following policies that reward and promote heroic imagination and action.

• • •

When healing businesses grow, it is not like spreading an empire. When healing businesses grow, they embrace, uplift, and liberate people from the trauma they have long endured.

When healing businesses grow, it is like spreading a ministry.

Consider the case of an empire that transformed into a ministry.

ASHOKA: FROM HELL TO HEAVEN

Unilever's evolution from an empire into a Healing Organization has a precedent that goes back 2,300 years to the kingdom of Ashoka, an emperor of the Maurya Dynasty who ruled almost all of the Indian subcontinent from 268 to 232 BCE. Grandson of Chandragupta Maurya, Ashoka did what he was expected to do and expanded the empire to stretch from present-day Afghanistan in the west to Bangladesh in the east and much of the rest of India.[27]

As a young man, Ashoka was an ill-tempered and nasty character, known for his ruthlessness and cruelty. He burned his stepbrother alive to consolidate his hold on power and built an elaborate torture chamber for his enemies that became known as Ashoka's Hell. In his lust to expand his sprawling empire ever further, he waged a massive war against the neighboring state of Kalinga in the northeastern part of India, a region that his grandfather had attempted to conquer without success. Hundreds of thousands were butchered in the fighting. In the final battle, Ashoka's army achieved complete victory.

After the victory, the emperor walked the battlefield to savor his triumph. But instead of experiencing his usual satisfaction at the screams of the wounded and the piles of enemy corpses, Ashoka began to experience a different kind of feeling stir in his heart. As he walked with his entourage down to the river nearby and saw that it flowed red from the blood that had been spilled in the battle, he underwent an awakening of conscience and consciousness. Ashoka realized in that moment what sages from every tradition know: All life is interconnected. Masters of all the world's wisdom lineages teach nonviolence, mercy, and compassion because they understand experientially that all life is one. Ashoka's enlightenment was commemorated with the renaming of that river, known to this day as Daya, or River of Mercy.

From that day on, Ashoka renounced war and violence, and preached and practiced compassion and kindness instead. In the succeeding months and years, he issued a large number of edicts, which were carved into stones all throughout the massive empire. Edict 13, an ancient precursor of the Truth and Reconciliation initiative that was so successful in South Africa, expresses the deep remorse he felt after observing the battle of Kalinga. The edict acknowledges:

> The conquest of a country previously unconquered involves the slaughter, death, and carrying away of the captive people. . . . This is a matter of profound sorrow and regret to His Sacred Majesty.[28]

Embracing the teachings of the Buddha, Ashoka became a passionate advocate of the doctrine of compassion. He sent missionaries far and wide, including his own son and daughter who established Buddhism in Ceylon (now Sri Lanka), off the southern tip of India.

Ashoka's edicts focused on moral precepts such as truthfulness, purity, mercy, and charity. His edict *On Benevolence* commanded those in power to use their influence to make life better for all. He scrapped his torture chamber and became a champion of fairness in the exercise of justice and kindness to prisoners. Ashoka eliminated slavery in his realm. His reverence for life extended to animals; he abolished the Royal Hunt and the practice of animal sacrifice and promoted vegetarianism. He is believed to be the first major ruler to have cared so deeply about animals and his rule has been described as "one of the very few instances in world history of a government treating its animals as citizens who are as deserving of his protection as the human residents."[29]

Ashoka's influence lives on, not only in the Buddhist legacy he left to much of Asia but as an inspiration to contemporary leaders. Novelist, historian, and futurist H. G. Wells (1866–1946) considered him to be one of the most exemplary rulers who ever lived. In *The Outline of History*, Wells wrote, "Amidst the tens of thousands of names of monarchs that crowd the columns of history, their majesties and graciousness and serenities and Royal highnesses and the like, the name of Ashoka shines, and shines, almost alone, a star."[30]

Ashoka is a singular example of the awakening of conscience and the transformation of an empire into a kind of ministry. From an army of soldiers bent on destruction and conquest, he created legions of missionaries spreading a benevolent message of peace and nonviolence.

BEYOND BUSINESS AS EMPIRE

With the demise of the British Empire and European colonialism in the mid-twentieth century, and the defeat of the Third Reich, followed by the disintegration of the Soviet Union, empire building has, for the moment, become less a matter of national or tribal ambition and more about global corporate activity. The same energy that drove the creation of military and political empires has been present in the world of business since the dawn of the era of modern capitalism in the late eighteenth century.

Driven by the philosophy of Social Darwinism as espoused by Herbert Spencer, Carnegie and other US industrial titans did their best to create monopolistic empires by manipulating markets, utilizing private armies to enforce their will. They didn't rely primarily on violence, and they couldn't actually enslave people, but they did their best to squeeze their workers and crush the nascent labor movement. They believed that the Darwinian theory of survival of the fittest could be applied to society so that "humanitarian impulses had to be resisted as nothing should be allowed to interfere with nature's laws, including the social struggle for existence."[31] Courted as a friend by Carnegie, Spencer was invited to visit the mills in Pittsburgh, which were meant to be an embodiment of the ideas in Spencer's books. At the end of his visit, Spencer reported, "Six months residence here would justify suicide."[32]

Carnegie was a complex character: despite his flaws, he was ultimately attempting to make the world better and he encouraged others to do the same. To paraphrase his famous Dictum, Carnegie advised:

- In the first third of life, get all the education you can.
- In the next third, make all the money you can.
- And finally, in the last third of life, give it all away to benefit humanity.

Carnegie, Rockefeller, Vanderbilt, and others gave away fortunes and endowed many wonderful cultural and charitable endeavors. They operated under certain assumptions that prevented them from considering a more humane approach to generating the wealth that they ultimately gave away. *Now we know that it's not only possible to generate wealth by benefiting all stakeholders, it is actually a more profitable path in the long run.*

TRULY HUMAN LEADERSHIP

The Healing Organization is a book about everyday heroes and how you can become one. One of the heroes who inspired the ideal of the Healing Organization is Bob Chapman.

After completing his MBA at the University of Michigan, Bob was offered a job by his father, Bill, who had acquired a controlling interest in an eighty-five-year-old company that made equipment for the brewing industry. The company, called Barry-Wehmiller, was struggling, and father and son collaborated to help save it. A few years in, when Bob was not yet thirty, his father died suddenly of a heart attack. Bob was now responsible for a money-losing company that did not appear to have much of a future. Fearing that the inexperienced young CEO would not be able to run the business, the banks called his loans.

In his book, *Everybody Matters*, coauthored with Raj Sisodia, Chapman shares the inspiring story of how he rose from those precarious beginnings to lead a company with nearly $3 billion in revenues in 2018, whose share price had grown at a compounded rate of 17 percent for twenty years. At last count, Bob had acquired more than a hundred companies and never sold one. Most of those were struggling or dying businesses, initially concentrated in small industrial towns in Wisconsin, Ohio, and Pennsylvania. Many of those towns were completely reliant on those businesses and facing disaster. One by one, Bob began to revitalize them.

How did he achieve such spectacular and unexpected success turning around these relatively low-tech manufacturing businesses—for example, building machines that make corrugated boxes or toilet paper—that have for decades been closing in the United States and moving to cheaper labor markets in Brazil or China?

The answer is that Bob Chapman does not so much acquire companies as *adopt* them. He doesn't have them dissected in order to sell their organs to the highest bidder. He embraces and nurtures them.

His "truly human leadership" formula is simple and is expressed in a statement prominently displayed on the wall at the company's headquarters in St. Louis:

We measure success by the way we touch the lives of people.

Not by power. Not by position. Not by money. Not by growth. Truly human leadership measures success by the way human lives are enriched and healed.

Bob evolved his approach through a series of epiphanies that awakened his conscience and opened his heart. The first one happened at church. Inspired and uplifted by his preacher's sermon, Bob began to consider if work could have a similar effect on people's souls. In that moment he knew: "Business can help heal what is broken in our society."

Bob also saw that many workers were playful and lively before they got to work and after they were off the job but depressed and grim as soon as they punched in. His second epiphany was that work could be fun. He began creating all kinds of enjoyable and playful activities that transformed the cultures of his companies.

The third epiphany, and one that has become talismanic at Barry-Wehmiller, happened as Bob empathized with a friend who was giving his daughter away at her wedding. He sensed the profound love his friend felt for his precious child and then he was overwhelmed with a powerful understanding: "Every one of the seven thousand people who work for us is *somebody's* precious child. Why are they not deserving of the same level of care and concern that I would have for my friend's daughter or that he would have for my son?" From that day on, the wedding story has become an essential part of the lore at the company and the most powerful symbol of its culture. In other words: It's not just business, it's personal.

Bob believes that he has been touched by grace, and he feels a deep sense of duty to share the wisdom he's received. "I didn't learn this at business school and I didn't get this from a business book. I can only say that these revelations were inspired by some higher power, because there is no way I thought of those things. I have an over-whelming feeling that I was blessed with a message that could pro-foundly change the world. When you feel that calling, you cannot say no to it."

Now in his early seventies, Bob travels nearly nonstop in his cor-porate jet, spreading his gospel as he scours the world looking for new businesses to buy and communities to heal. When we asked him why he continues working when he could be relaxing and enjoying the fruits of his success and his (at last count) twenty-six grandchil-dren, he replied, "On my deathbed, I will not be proud of the ma-chines we built or the money we made. I will be proud of the lives we touched. And I want to touch as many lives as I can while I'm here."

Bob Chapman is not growing a business empire; he is spreading a healing ministry. When companies like 3G Capital grow, life gets worse for more and more people. When healing companies like Barry-Wehmiller grow, life gets better for all who enter its embrace. When 3G Capital comes exploring whether it should buy your busi-ness, it casts a pall over the whole enterprise and creates a funereal atmosphere. But when Bob Chapman comes to look at a potential adoption, people's hearts are filled with hope.

What Bob Chapman has created at Barry-Wehmiller has touched and significantly altered the trajectory of many thousands of lives. Randall Fleming is an example. For many years, he was known as Randy—a tall, intimidating Navy veteran who worked as a fabricator in the machine shop. He recalls, "I was diagnosed with two emotions when I had to see a counselor as part of my divorce: I was either mad or angry. I didn't really have anything else."

Randy was frustrated in his work and felt unappreciated. He had not received a raise in five years because he had topped out as a fab-ricator. He was cynical when he first heard about the Barry-Wehmiller approach to continuous improvement and cultural transformation known as "L3" (Living Legacy of Leadership). But L3 leader Ken Coppens kept coming around to speak with Randy, asking him what he really wanted to be doing at the company. Slowly, Randy realized that this concern for his well-being was genuine.

One of the unique things about Barry-Wehmiller is its embrace of "courageous patience." Jim Collins popularized the bus metaphor: Get the right people on the bus, and make sure they are in the right seats. Bob Chapman thinks of it differently. To him, it is about having a safe bus and a driver who knows where we are going and how to get us there. Everybody gets to go on the bus. And if you miss it the first time, the bus will come around again. And again. And, if need be, again. There is no pressure for people to "get it" right away; they will get it when they get it, and the bus will be back.

That is what happened with Randy. It turned out there was an opening in the L3 team for somebody from manufacturing and Randy was offered the role. This began his transformation into a different kind of person, who now insists that people called him Randall instead of Randy.

The most gratifying part of Randall's journey is the healing of his relationship with his daughters. He explains, "My daughters grew up around this big, mean, scary guy." But, as he began to shed his armor and open his heart, his daughters noticed and responded. Randall's joy is palpable as he exults, "Now, they are my best friends, and they tell me I am their best friend. Whenever anything happens in their life, they call me first, and I call them every week. We talk about how things are going in their life. It is a complete one-eighty from before."

How many generations of the Fleming family are being healed by Randall's experience at Barry-Wehmiller?

The people who work at Barry-Wehmiller are, like Randall, predominantly middle-aged, blue-collar men with high school educations or less. But walk into a room full of such men and ask a simple question: "Tell me how it used to be before this company got acquired by B-W, and how things changed?" You will be moved by what happens next. Burly men get choked up and a number soon have tears streaming down their faces as they speak from the heart about the profound difference the company has made in their lives.

There is something powerful at work here, and it is rooted in the simplest and yet most profound aspects of what it means to be a human being: We all want to know that who we are and what we do matters. We all want to be treated with care and respect. We want to feel safe, and we want to know that our children will be all right.

This is the priceless gift that Bob Chapman has given so many.

STOP PULLING THE SHOCK LEVER: IT'S TIME TO HEAL

Think about what your business would feel like if it cared for people's souls as much as it cares about profit. Many businesses start out with the intention of being ministries, but eventually start behaving more like empires. When Sam Walton was alive, Walmart was guided by a spirit of taking care of people in small-town America. After his passing, the company became a voracious growth machine and lost its connection to caring. It became an empire. The company is now trying to recapture the healing energy with which it was originally endowed.

Home Depot was a caring and flourishing organization until Bob Nardelli came in with a dehumanizing, cost-cutting approach that nearly ruined the business. Starbucks began as a healing organization under Howard Schultz's leadership but lost its higher purpose when his successor came in and focused solely on growth. It revitalized its healing vision when Schultz came back to lead the company.

Now is the time to incorporate healing into all aspects of corporate life.

It's time to renounce the destructive legacy of Attila the Hun, Chainsaw Al, and Tyrannosaurus Lemann.

It's time to stop pulling the shock lever when the person in the white coat says, "Continue the experiment."

It's time to reject the idea that it's OK to be "merely a little cog in the machinery" that pollutes the planet and makes lives miserable.

It's time to stop inflicting pain and suffering on ourselves and others.

It is time to heal.

• • •

In the next section, we share stories of healing companies that can help us all realize what's possible.

PART 2

THE JOY THAT
IS POSSIBLE:
STORIES OF
ORGANIZATIONS
THAT HEAL

THE POWER OF INNOCENCE

HOW JAIPUR RUGS BROUGHT DIGNITY, PROSPERITY, AND HOPE TO 40,000 LOWER-CASTE WOMEN IN INDIA

Tourist guidebooks to Jaipur, India's magnificent "Pink City," renowned for its historic palaces and princely heritage, warn: *Do not visit during the extremely hot summer months of March through July, when daytime temperatures often exceed 100 degrees and the dust and pollution combine to make Jaipur unbearable.*

Although for tourists the summer weather may be unbearable, it's just normal for people who were raised in these conditions. It is also normal for lower-caste workers, especially women, to work in conditions that are much more oppressive than the summer heat.

India's ancient caste system is a rigid form of social stratification with thousands of distinctions and sub-castes based on inherited and unchangeable assignment to different types of work. There are four main tiers: At the top are the priestly, scholarly Brahmins, and below them the warrior and administrative class. The next level down are traders, merchants, and business professionals, and at the bottom the Shudras, or menial workers, a category that includes the so-called "untouchables," who for centuries have done the dirtiest and most toxic jobs like sanitation, leather tanning, and funeral work. Born into a life of punishing and poorly paid labor, members of the lowest caste are considered unclean and are prohibited from touching the skin or even the clothing of any of the "higher" castes. The prejudice was so extreme that in some regions of India members of the lowest castes were forced to wear bells to warn of their proximity as it was believed that even their shadows were contaminating. Upper castes had special cleansing ritual baths to restore their purity after even incidental exposure, and untouchables were subject to severe beatings if they inadvertently touched—or even cast a shadow—on a member of a higher caste.

Although discrimination of this kind is now officially illegal, the tradition remains widespread and it is especially burdensome for women.

A girl born into an untouchable family occupies the lowest of the low tiers of social status in India. These girls are fed last, educated minimally if at all, and conscripted into housework at an early age, including childcare for their younger siblings. Married off in their early teens, they are sent to another family and a lifetime of servitude. They are expected to bear and care for several children, cook and clean for the household, serve their mothers-in-law and husbands—and carry on their hereditary occupation, earning money that will be handed to the mother-in-law. Alcoholism is rampant among lower-caste men, and verbal and physical abuse of women is commonplace.

Abuse is also rampant in the workplace. Women weavers, for example, are frequently exploited by middlemen from the merchant class; they take orders for carpets and then find female weavers in villages to do the specified work on a strict deadline for a fraction of what the middlemen receive. It's been a regular practice for these operators to short-change the women, using insignificant mistakes or tiny blemishes to justify withholding half or more of the agreed-upon amount. In many cases the women are not paid at all and have very little recourse.

Jaipur Rugs is an organization devoted to healing this tragic legacy.

VISITING JAIPUR RUGS IN JULY

It was indeed hot in early July as we drove from Jaipur toward the village of Manpura through the arid, semi-desert landscape that is typical of the region. The early monsoon rains of mid-June had given rise to a pleasing amount of greenery. In the car was Mr. Nand Kishore Chaudhary (affectionately known as NKC) the visionary founder of Jaipur Rugs. A modest man with an open face and sweet smile, NKC spoke in mesmerizing fashion about launching his widely admired and highly successful company. As he spoke, it became clear that beneath his humble demeanor was a steely determination and an abiding passion to address the plight of the weavers.

As we pulled off the paved road onto the bumpy, dirt access lane into the tiny village, one of hundreds in five Indian states where some of the company's 40,000 weavers live and work, we passed stray dogs and clusters of men sitting idly at tea stalls, staring at us as we drove by them trailing clouds of dust. Finally, we pulled up to a modest house that was covered with a slanted tin roof and entered its court-yard. There stood several rustic-looking looms arranged parallel to one another, about four feet apart, stretching from the house to the courtyard wall.

At each of the looms, two women wearing brightly colored sarees squatted side by side on a low wooden platform, facing an arrangement of vertical threads stretching from a wooden bar about seven feet up to another bar at ground level. Behind them lay assorted bundles of colored yarn to be woven into the prescribed design. About six inches or so of a long printed pattern that carried the blueprint for the carpet being created was visible in front of the women, just below the line they were weaving. Using a tiny sickle in one hand, they tied the individual knots that created each miniscule piece of the carpet. The women's hands flew back and forth with astonishing dexterity.

They laughed and talked almost as fast as they wove. One woman worked with a small child draped across her lap. It was evident immediately that this work required great coordination, stamina, skill, and precision, but these women made it look easy. How could they squat like that for hours at a time doing this painstaking task while exuding so much joy?

During our drive, NKC prepared us for what we were witnessing by explaining: "These women are natural leaders, with tremendous problem-solving skills. They possess strength and ingenuity that we can only marvel at."

One such woman is Shanti.

SHANTI'S STORY

Ten years ago, Shanti's work as a weaver was a daily struggle. She worked long hours under terrible conditions and dealt regularly with unscrupulous middlemen. Nevertheless, through her exceptional determination and skill she had become a supervisor of other weavers and their representative in dealing with the contractors.

One day, one of the seemingly more trustworthy middlemen asked her for a short-term loan to pay his taxes. In good faith, Shanti agreed, using money she had earmarked for her workers (approximately Rs. 3,500, about $70 at the time). The middleman absconded with the money and it became clear that she wouldn't get it back. At the same time, other middlemen all reneged on promised payments based on flimsy excuses, and the rug work in her village came to a standstill.

Shanti sold her jewelry—all she had left of value—to raise money to pay her workers. With no income and no support from her in-laws (who were infuriated by her sale of the jewelry), and with a husband who drank to excess and shirked responsibility, her situation was dire. At one point she was compelled to send her eight-year-old daughter to beg the local shopkeeper for some flour so that she could provide minimal sustenance for her family.

Desperate for work, Shanti traveled to a nearby village to do menial labor for a cement company. While there, she overheard a few women talking about a man named Harphool who was offering a weaving job that would pay Rs. 100 ($2) a day, much more than she could earn at the cement factory. Shanti got his phone number and called him immediately. Although she didn't know it at the time, Harphool was a distant relative, and he invited her to come to his office. He offered her a loom, and eventually as he observed her ability both to weave and to bring out the best in other weavers, he asked Shanti to add more looms. She was hesitant about the extra responsibility as she had six children—five girls and a boy. But there was something different about Harphool, and despite her previous betrayals and disappointments at the hands of brokers, she trusted him.

Today, Shanti directly oversees six looms and her village has a total of forty-two, with more than a hundred artisans. She leads all the weavers and continues to weave herself. Increasingly, she also interacts with customers, welcoming them to her village to experience the magic of Jaipur Rugs.

When we arrived, she greeted us with a hearty "How are you?" in heavily accented English, much to the amusement of the other women. Shanti offered us tea, and motioned for us to walk a few feet toward the clean, simple house, where a few chairs had been set up in a living room next to a loud, clanging air cooler. A few of the other women joined us as we sat and talked for the next hour. Also with us were Harphool, who now manages the region for Jaipur Rugs, and NKC's daughter Kavita, who is head of design for the company.

Previously, all of these women would have covered their faces completely in front of others, and would've presented themselves with apologetic body language. But here they were upright, heads uncovered, greeting us with wide smiles and sparkling eyes.

When asked to share about her experience, Shanti explains, "When I joined, I was shy, under-confident, and lacked knowledge, so much so that I did not even dare look at Bhaisaab [which means *older brother*, referring to NKC] when he paid us a visit. But today, I can talk to any person confidently."

Shanti is most delighted about the effect her transformation has had on her family. Her husband is in recovery from his drinking problem. Her oldest daughter is now happily married and in college. The next one is three years younger and also in college. The next two are in the eleventh and tenth grade and thriving academically. The youngest girl is in the sixth grade and her son is in the second grade. Shanti is determined that all six of her children will complete college, something until recently unthinkable for an untouchable.

Thanks to the Jaipur Rugs Foundation that provides an adult education program in the villages it serves, Shanti's own education has come a long way since she was forced to leave school in first grade. She enthuses, "I have learned so much. This has changed the way my family and I are able to live."

Shanti is devoted to supporting and uplifting the women who work with her and encourages them to take advantage of the educational opportunities provided by the foundation. She is beloved by her workers and their families. The villagers call her *Devi*, the Hindi word for *Goddess*. Shanti's naturally loving and generous nature has blossomed as her self-confidence has grown. Her growing ability to inspire others is, she believes, an effect of the culture of Jaipur Rugs, and the way it was represented first by Harphool and then by NKC himself. They inspired her to believe in herself and to stand up for herself when necessary. She exclaims, "I naturally think of what other people need before I think of myself. If I'm hungry, I'll first offer food to someone who is in need. I can't bear to see anyone unhappy. But I have also learned to believe in myself. Now, whenever I have a contradictory opinion, even if the other person may not appreciate it, I tell them. I speak up."

When asked about her dreams for where she wants to be five years from now, Shanti says, "I have come pretty far working here and I think I can go farther." She explained that many people move away from the village looking for work, so she proposed to NKC the idea of a new center to provide expanded training, education, and employment to keep the village community strong.

How was an uneducated, impoverished, abused woman transformed into a happy, successful, and fulfilled entrepreneur? Shanti's distant cousin Harphool played a key role. His name means "one who laughs and cheers people up," and it is apt.

HARPHOOL'S STORY

Harphool came from a poor, debt-ridden family, with five siblings and an alcoholic father. Like Shanti, his sisters were denied elementary education and hired out to contractors to work on looms where they were often harassed and abused. His parents tolerated the mistreatment because they needed money desperately. When Harphool was in the fourth grade, he decided to try working on a loom himself to help feed his family and to protect his sisters.

Despite sustaining a severe injury at work, he continued on the looms to ensure that his sisters weren't ill-treated. Eventually he found his way to a weaving group run by a good man who taught him new techniques, and introduced him to considerations of design, quality, etc. NKC visited this facility and, impressed with Harphool's work, asked him to become part of Jaipur Rugs.

NKC brought Harphool to the headquarters in Jaipur and personally showed him around, then asked him to be a supervisor. Harphool couldn't believe it and protested that nobody would pay heed to "a poor and backward soul" like himself. But NKC saw his potential and told him that an honest, sincere, and innocent man needn't ever be afraid, and that determination woven together with goodness would yield success. Soon, Harphool was not only a supervisor but also one of the leading recruiters of weavers for the company. Within six months, he had brought seventy-nine new looms into the Jaipur Rugs family, handling all the arrangements. Harphool explains that NKCs presence and encouragement helped him achieve things he previously considered unimaginable.

With tears of gratitude welling in his eyes, Harphool recounts, "My body may be a gift of the almighty, but Bhaisaab gave me my soul. The value and reputation I have today is all because of him. Today, our extended family is strong. Our kids are all studying in good schools and doing well. . . . Our lives are completely transformed."

Shanti expresses her love for Harphool "He motivates me . . . whenever I see him, I feel like bettering myself."

And for NKC, "He spends time with us, eats with us, and never has he treated us like untouchables. And seeing that, all of us have started treating each other as equals. See, even she [pointing to Kavita, daughter of NKC] is sitting on the floor with us. That's how they treat us."

She concludes, "He cares for us more than our parents ever could. He is my god."

SO WHO IS NAND KISHORE CHAUDHARY?

He's one of the most unusual CEOs in the world. Known as the "Gandhi of the carpet industry," he is an indefatigable champion of those on the lowest rung of Indian society. Compelled by his passion to generate a "first chance" for those who have been denied opportunity, he is an embodiment of moral courage and the creator of a truly healing organization.

Born in 1953 in the Shekhawati region of Rajasthan, NKC was one of six children. His father ran a shoe store and his mother was a housewife. His greatest influences from childhood were the inspiration he received from contemplating the beauty of nature, and the wisdom he gained from reading great works such as Gandhi's *Autobiography: The Story of My Experiments with Truth, The Bhagavad Gita*, and the poetry of Rabindranath Tagore. His soul was infused with a love of wisdom, beauty, art, and compassion for others.

After graduating college he struggled to find his place: He didn't feel drawn to work in his father's shoe store and soon became disenchanted with a job in a bank. In 1977, NKC married Sulochana and they had three daughters, followed by two sons.

A turning point in his life was his meeting with British adventurer, photojournalist, and art historian Ilay Cooper—author of many books, including *The Painted Towns of Shekhawati*. The men bonded over their mutual love of the local artisanal traditions and have been friends ever since. Cooper supported NKC's intuition that he needed to find his own path and that it would involve making the world more beautiful. Cooper also inspired NKC to evolve his attitude about the potential of women. NKC's first three children were all female, and in India, as in many other places, the lack of a male heir is considered a failure. But Cooper advised him: "You should bring up your daughters in an open and friendly environment. Give them all the opportunities and support they need to grow. There should be no difference in their upbringing just because of their gender."

NKC heeded his friend's counsel in a way that not only brought out the best in his daughters but ultimately in his business and the communities he serves.

Inspired by the beautiful patterns it contained, he learned the ancient art of carpet weaving and started teaching it to people in local tribal communities. With a small loan from his father, he began in Rajasthan in 1978 with just two looms. He has grown the company to more than seven thousand looms in five Indian states, with total annual revenues of approximately $20 million. He has been honored by many global business groups including Ernst & Young, which named him Entrepreneur of the Year in 2010.

THE MASTER KEY: INNOCENCE

How did NKC create such an extraordinary healing organization? People who know him from childhood say that he always possessed a quality best described as *innocence*. Friends from college also remember him as a pure spirit, and say that he was "the same outside as he was inside." This was seen as a positive and endearing quality, until he decided to go into business. Then, many people told him he wouldn't be able to succeed because he wasn't hardened and cunning like other businessmen. There is a common expression in India: "A good person is a bad businessman."

But others observed his way of being and predicted that he would be successful, because word would get around that he was a genuinely caring and incorruptible character. They saw that his innocence would become his greatest strength. NKC reflects, "Today, all the cunning people who used only their minds to get ahead in the carpet industry have fallen behind me. They used their cleverness to figure out how to exploit workers, cheat customers, not pay weavers what they were owed. Because I was innocent, I never got trapped into doing any of those things. I just did what I felt was right. It was not difficult to see what that was." For NKC, it was self-evident that the standard industry practices of using toxic dyes, child labor, and the exploitation and abuse of the weavers were unconscionable. He changed all that from the beginning and magnetized other good people to follow.

For NKC, innocence is the master key to healing. He explains, "My life's purpose is to create a business that is all about innocence: serving innocents, employing innocents, led by innocents. Innocent people have the power to see and understand things others cannot. . . . That will transform the business beyond recognition, because they can go so deep into the human condition."

His daughter Kavita adds, "Innocents are in closer touch with the flow of life. They are more natural, not forced. We see this in our designers. There is less ego, less self-consciousness. There is a kind of auspiciousness. Innocents are able to sense global trends before they become apparent to others. They are more finely tuned in to what is emerging."

There is a kind of innocence that is naive and weak and helpless, the innocence of a young child. There is also the innocence on the other side of wholeness, a mature, loving *chosen* innocence. For NKC, innocents have extraordinary power. Theirs is a power *with* others, not power *over* others. Innocents never knowingly cause others to suffer. They are, in Franciscan terms, *instruments of peace*. We need more innocent leaders in the world, in every sphere: politics, business, civil society. Like NKC, such leaders have a purity of heart and intent, supported by a spine of steel.

THE HIGHER SCHOOL OF UNLEARNING

When he started, NKC was primarily focused on quality and production, but knew nothing about marketing, finance, and HR. Initially, he lost a lot of money. But his innocence and "beginner's mind" made him humble and agile and enabled him to adapt and develop an effective system for integrating these essential business functions.

His commitment to people and culture was challenged when his company started to grow rapidly. He engaged well-educated business professionals to help manage his network of uneducated artisans, but quickly realized that managers who didn't empathize with his weavers would destroy the essence of his creation.

NKS developed a unique management training approach that he calls the "Higher School of Unlearning." Aspiring leaders at Jaipur Rugs all learn to weave; they experience every stage of the generation of a carpet and learn to empathize with workers at every step of the supply chain.

The Higher School of Unlearning is based on one of NKC's favorite adages: *finding yourself through losing yourself.*

In other words, lose your egocentrism to discover your essence. He explains, "When the ego is in charge, things don't work so well—it leads to cunningness, defensiveness, the need to prove oneself right." Instead, he counsels that in the search for solutions we open the heart and access more intuitive wisdom.

This approach is called "Mindful Leadership" in contemporary parlance, but that term didn't exist when he began practicing it.[1] He says, "If you are focused on the goal and the desire to acquire something, you will be in a state of tension. Your purpose is to achieve the goal rather than do the work. The results may come, but not the grace. What you work on in the present defines what the future will be. If you work in the present without the desire to attain something, what you get then is a state of grace. From grace comes gratitude."

NKC believes that many business leaders need to shed the fear and inferiority complex that drives greed, control, unethical practices, and undue haste. Unlearning fear liberates natural compassion. As NKC states, "Leaders driven by love will bring sustainability and healing to the business, as well as for themselves. When you come from a place of love, you will go deep and not be driven by external competition. Then you don't need to search for the market; it will find you."

HEALING CUSTOMERS

The market is finding Jaipur Rugs and the company is mobilizing to keep up with the demand. This is a reflection of NKC's deep understanding of what his customers really need. In response to the fundamental question "Why do people buy things?" he offers a profound insight: "There is some empty space inside them and they are trying to fill it. It gives them some excitement, the thrill of something new. But the more you buy, the more that empty space keeps expanding. *The only way to fill that space is through love and healing.*"

He adds, "When business can be the conduit for people to connect with each other, it will lead to true satisfaction. The market will move toward the essential things that customers truly need."

In the last decade, the market has moved in just this way. Jaipur Rugs is receiving international acclaim. Business leaders from all over the world, including Unilever CEO Paul Polman and senior executives from Ernst & Young and Bain Consulting, are making the pilgrimage to Jaipur to attempt to understand the magic. Visitors are all shown how to weave, and each creates a loop or two for a carpet that will ultimately be an expression of the curiosity, energy, and love of Jaipur Rugs' global family. The rug is also a living symbol of a corporate culture that is predicated on a systematic approach to the welfare of all its stakeholders, beginning with the artisans. As the company has a positive impact on the lives of weavers, they are increasingly connected with customers, and customers experience this as something special.

When producers and customers connect in this way, there's a healing effect and a flow of abundance. In NKC's words, "When we tell our story to customers, they are not only willing but happy to pay more than before. They tell us that there is no better company, no more honest people with higher integrity than they have encountered here. Those customers are being healed as well."

Customers (mostly interior designers) come to Jaipur, get to know the weavers, and leave feeling intimately connected with the company and especially with the weavers. When they hear the stories of the weavers and how their lives have been transformed by being part of the Jaipur Rugs family, they are deeply moved. They come to understand and appreciate what truly goes into producing a carpet: how much hard work, care, diligence, and—in the past and with other carpet companies today—how much suffering. Customers come to realize that they can be part of a healing transformation of these lives and future generations.

As NKC expresses it: "We don't sell carpets, we offer a family's blessings."

Jaipur Rugs' healing culture has also earned it a strong reputation with retailers. NKC told us, "We learned recently that one of our competitors reached out to one of our major retail partners, offering to copy our unique designs and undercut our prices. They are driven by fear and an inferiority complex. But our retail partner refused to do business with them, considering what they are doing unethical. I am confident we will be the most competitive company in the industry, in addition to being the most joyful!"

NKC sees the potential for more transformation to come. "I feel that our way of operating will grow within our industry as well as in other industries."

•   •   •

JAIPUR RUGS IS A BUSINESS *AND* AN ASHRAM

A firm believer in the power of business to transform society, NKC states, "I have always been an advocate of for-profit solutions to social issues. My philosophy is this: Give people a way to make a living, not just charity. In this way, your efforts are sustainable, and so are the livelihoods of all the people you touch."

He adds, "Governments and NGOs have tried very hard, but ultimately it is up to business to heal society. Governments often give people money without them having to work for it. I think about how business can transform society. Companies understand how to operate, how to serve customers, how to make a profit. But if a business is driven by love, it can truly transform. People want to give and receive love. Businesses that enable that will not only succeed, they will also heal society."

•   •   •

We began by sharing the Jaipur Rugs story because the depth of suffering previously endured by its workers, living in extreme poverty in rural India, exceeds anything experienced by the blue-collar workers of small-town America. Shanti and Harphool would consider out-of-work American coal miners and factory workers to be wealthy and privileged. Their desperation was far greater than most can appreciate and their lives have been fundamentally transformed by a Healing Organization. If this can happen in rural India, it can also happen in America and around the world. It is happening! Let's explore some more notable examples.

THE UTILITY
OF LOVE

HOW DTE ENERGY SAVED ITSELF AFTER THE
FINANCIAL CRISIS AND THEN HELPED SAVE DETROIT

On November 4, 1879, Thomas Edison applied to patent the light bulb, effectively launching the age of electricity. In 1886, the Edison Illuminating Company of Detroit was founded, which later evolved into Detroit Edison. In 1937, the Detroit City Gas Company merged with several other companies to form Michigan Consolidated Gas Company. The gas and electric companies came together in 1996 to form DTE Energy, with a vision to offer integrated energy solutions. Today, DTE's 10,500 employees serve 450 Michigan communities.

Few businesses are as tied to their communities or have as much invested in them as electric and gas utilities. While they are private or publicly traded corporations, historically most operated as "natural monopolies," since it did not make sense to have multiple companies install electric poles and run wires and gas pipelines everywhere. The companies were regulated and operated under a "universal service" mandate, and most were not known for innovation or superior customer care. That started to change around 1990, as competition and customer choice were introduced into the industry. Many utilities started to evolve and innovate in ways they hadn't done before.

We spoke at length with Gerry Anderson, chairman and CEO of DTE Energy, about how his company has become a Healing Organization.

When Gerry became president of DTE Energy in 2004, he asked himself and his leadership team a challenging question: "Is this a good company?"

The answer was clear: no.

Customers weren't happy: J. D. Power assessments rated DTE's customer satisfaction level near the bottom of the eighteen companies in their peer group.

Employees weren't happy: Internal Gallup surveys demonstrated bottom quartile employee engagement scores year after year.

Shareholders weren't happy: Total shareholder returns were consistently in the bottom quartile of performance metrics.

And the unhappiness of customers, employees, and shareholders was, unsurprisingly, mirrored in operating metrics that benchmarked poorly on cost and reliability.

In his new leadership position, Gerry began to feel responsible, for the first time, for the whole system that was generating unhappiness and poor performance. He states, "I looked around at all of this and thought, 'I am now president of this operation and the picture is not a pretty one.'"

Change didn't happen immediately. Gerry and his team grappled with the challenge and eventually concluded that all of these problems were a consequence of a "culture of mediocrity."

As a former McKinsey consultant, Gerry's initial approach to turning the company around was to drive better performance metrics in a project-oriented way. While operational numbers began to improve with this effort, employee engagement levels actually got worse.

Gerry had the first of a series of epiphanies that changed his life and, ultimately, the lives of thousands of others in the greater Detroit area. It dawned on him that, although costs and reliability did improve somewhat through the performance initiative, the core issue of culture wasn't being addressed; if anything, the culture was being harmed. He stated: "People hated it because it felt imposed, top-down. . . . Our already mediocre employee engagement level went lower!"

He began searching for a better approach and found his way to the idea of continuous improvement. The heart of successful continuous improvement programs is an unwavering commitment to the value and well-being of people. Gerry mused, "Perhaps this was the antidote to our cultural ailment?"

He reflected deeply and realized that any equivocation or misaligned action would undermine the effort. He realized: "When you put your people at risk as you pursue productivity and improved quality, it is a very difficult thing then to ask them for their energy and their creativity to improve the company. It is just not a bargain people will strike. I realized that I needed to genuinely show people that they would come first, that we really do treasure them and care about them."

Gerry's evolving people-centered philosophy would soon be put to a severe test. "I was internalizing all this and had integrated it into our plan for 2008. Then the financial crisis hit and the world fell apart. Everything in Detroit just melted down." He continued, "We were being talked about as potentially a junk bond utility that was surely going to be thrown on the scrap heap with the rest of Detroit."

Gerry remembers very vividly the period around Thanksgiving in 2008, when he challenged his leadership team to come back after the holiday with a plan to prevent the company from becoming junk. A couple of the senior leaders came to him and said, "Look, we've done the math. We think a couple of hundred million dollars just walked out the door. This *is* going to turn us into a junk bond utility. To offset it, we don't see any other way than big layoffs."

The first principle of the continuous improvement philosophy that he had promised to implement is, "Don't put your people at risk as the first move when your company faces a challenge." He realized that this was an inflection point in terms of people's trust in him and in the company. After intense deliberations, the leadership of DTE decided to share the stark truth with all of its employees while reiterating their commitment to people.

Gerry addressed the entire company: "I can't make promises about the ultimate outcome, but we will promise this: the last lever we will pull to keep this company healthy is a layoff." He added, "But for us to make good on this commitment, you are going to have to bring more energy, focus, and creativity to helping us fix this company and keeping it healthy than you've ever brought before. If you make good on that commitment, . . . there is a very good chance that we can get through this together."

He explains: "I shared the message with all ten thousand of our employees. People watched with intense interest because they were afraid as hell. They were burning to know what was going to happen and what it meant for them. It was not a fancy communication; actually when I look back on it, I realize it was fairly crude. But it didn't need to be fancy, it just needed to be honest."

What followed in 2009 surprised everyone, including Gerry: "Month after month, my controller would come to me and tell me that we had beat the plan. . . . I couldn't quite believe it! We headed into April, when GM declared bankruptcy, and then into May, when Chrysler was spiraling down. By the middle of the year, we were still beating our plan. I told the controller to check his financial close model, because it didn't make sense to me that, with all the turbulence and failure going on around us, we were still okay.

"Then we went through the summer and into July and August. It was unseasonably cool in August and I realized we wouldn't be selling much energy for cooling. I thought, 'This is really going to be a bad month.' But he came back in early September to give me the August results and again we beat our plan. I remember actually getting agitated, telling him, 'You're missing something; your model has to be broken.' He shouted in reply, 'There is nothing missing. It is just what is happening!'"

This is when he realized that DTE's people were bringing unprecedented energy, creativity, and discipline to their work. They were streamlining processes, making fewer costly mistakes, and working in a much more efficient manner. The culture had really changed, and Gerry told the board in a year-end meeting: "I learned an incredible lesson in leadership. I discovered what people are capable of when they really believe in something."

Gerry calls this his "year of lessons." He confesses that he didn't have a "model in his head" for this kind of organizational transformation, "for what it meant for a culture to come together and really perform at a different level, the way an orchestra or an athletic team can when they are really inspired. But I saw our company do it. I saw people behaving in nonlinear, unexpected, remarkably positive ways."

By the beginning of 2010, the worst of the crisis was over and DTE's business had stabilized. Gerry was promoted to CEO and was contemplating how to sustain and extend the cultural transformation that had allowed the company to thrive under duress.

He realized that interacting with and listening to his employees was a critical element in demonstrating his care and in deepening his understanding, so he continued a monthly practice of sharing breakfast with about a dozen employees that he had begun during the crisis.

After the first of these breakfasts in January 2010, a woman stayed behind. He asked what was on her mind and she responded: "I never got a chance to thank you for what happened last year, for what the company did." She continued, "My husband and I have three young kids. He worked at one of the auto companies and was laid off. I had no idea what would happen to us as a family if I had been laid off, too, and I was deathly afraid. When you told us that if we band together we will get through this, I took hope in that and it got us through a very dark time. I just wanted to thank you for that."

And then she said something that sparked Gerry's next epiphany: "But, you know, a lot of my friends are still not okay and our community is not okay. There's a lot around us that is still hurting. Can we do something to help?"

Gerry was moved that this woman had shifted so quickly to thinking about the suffering of others. He realized, "Our people's energy is turning from self-preservation to what they can do to help other people."

He shared the story with his leadership team and challenged them: "What we need to do now is turn our people's energies from saving ourselves to helping to save our community."

Detroit had been in decline for decades. The population had decreased from 1.85 million in 1950 to 677,000 in 2015, as its once-dominant auto industry struggled to compete and moved operations to other regions in search of lower costs. Murder and general crime rates were among the highest in the country, and much of the area was in severe urban decay. Trapped in a downward spiral, the struggling city absorbed additional blows as the Great Recession of 2008 hit, causing mass layoffs and bankruptcies among some of its largest remaining employers. The city itself filed for bankruptcy in 2013, marking the largest municipal bankruptcy case in US history.

Gerry and his team realized that DTE Energy could be a powerful catalyst for revitalization and healing in Detroit and throughout the state by promoting economic development in the framework of a people-centered, continuous improvement philosophy. DTE began developing and leading initiatives to improve the transit system, race relations, education, health care, and vocational training in the greater Detroit area. He recalls,

Our region had a real history and reputation for division across political boundaries and along racial lines. Our community needed healing. It needed prosperity, economic life. I assembled the leaders of the biggest players in the region—GM, Ford, DTE, Blue Cross, Penske, Quicken Loans, Little Caesars, the owners of the Detroit Tigers, the Detroit Red Wings, along with the heads of the two biggest foundations—and asked if they would help me with this. The answer I got was, "Yes, but there are other things in our community that need to be fixed, too. We have employment issues, workforce issues, education issues. We have transit issues that especially affect the poor in our community; they don't have the means to get from place to place to be educated or get to work. We need to work on all that." They also wanted to work on creating wonderful places for people to be in and create a better image of the community and attract people to want to live in the city again. So we started with a half a dozen of those sorts of priorities and we have been working on those ever since.

DTE began to talk about a dual mission, what they called their Aspiration: be the best-operated energy company in North America, and, in so doing, change the narrative about Michigan from mediocrity and despair to excellence and hope. As Gerry explains: "We committed to be a force for growth and prosperity in the communities where we live and work."

They were on fire with the realization of the vital importance of their work: Energy is fundamental; it flows through every facet of daily life. How it is generated and how it can be sustainable were issues of the highest importance for all their stakeholders. They became passionate about crafting a statement of purpose and building communications around it. Eventually they created this:

We serve with our energy, the lifeblood of communities, and the engine of progress.

Gerry reflects, "I wasn't sure how it would play out. I felt that this was incredibly inspiring, but wondered if it would be received that way—or if it was going to be an embarrassing flop that would be met with indifference."

They presented the purpose statement, accompanied by music and imagery, to all ten thousand of their people in a series of town halls. The response was anything but indifferent: People erupted with joy and gave a sustained standing ovation. Gerry recounts: "When we showed it in our power plants, employees who had worked at the company for decades were in tears. For the first time in their lives they felt recognized for the importance of what they do."

Connecting to a higher sense of purpose generated even more positive energy. And that was strengthened further when they reconsidered and redefined their values statement in a way that aligned with the purpose and deepened the sense of meaning for all their people.

So what was the consequence of all this healing love for DTE's performance?

Employee engagement, as measured by Gallup, used to be in the bottom quartile; union employees used to be bottom decile (the ninth percentile at one point). In the last three surveys, DTE has placed in the top 5 percent of Gallup's global database, and its union employees are at the ninetieth percentile. DTE has received Gallup's Great Workplace Award for seven years in a row—one of only eleven companies worldwide over several decades to accomplish that.

In 2017, DTE was rated number one by business customers, and second for residential customers, in the entire Midwest. DTE's total shareholder return for the decade that ran through the end of 2017 was 275 percent, a performance that placed the company in the top quartile. The total shareholder return for the S&P Utility Index over the period was around 83 percent.

Gerry concludes, "Our journey is living proof that for employees, customers, and communities, business can be a tremendously powerful healing force."

THE ZEN
OF BROWNIES

HOW GREYSTON BAKERY'S OPEN-HIRING POLICY IS HEALING THE CYCLE OF CRIME AND INCARCERATION

Let's begin this chapter by meditating on a Zen poem:

> Drink tea and savor life.
> With the first sip . . . joy
> With the second . . . peace
> With the third . . . a brownie.

Greyston Bakery in Yonkers, New York, bakes 35,000 pounds of brownies every day. For many years their delicious pies and cakes were served in many of Manhattan's finest restaurants and their famous brownies are the key ingredient in every pint of Ben & Jerry's iconic chocolate fudge brownie ice cream. Greyston is a profitable enterprise that was intentionally designed to heal the wounds of homelessness and hopelessness.

It is the fulfillment of the compassionate vision of rocket scientist turned Zen master Bernie Glassman (1939–2018). Glassman was a remarkable being who conceived and articulated a philosophy of the Healing Organization and then brought it to life.

Born in Brooklyn, New York, in 1939 to Jewish immigrant parents from Eastern Europe, he was a gifted student with a wide range of interests. After graduating from Brooklyn Polytechnic Institute he got a job at McDonnell Douglas as an aeronautical engineer. Ten years later, in 1970, he completed a PhD in applied mathematics from UCLA.

How did this rocket scientist and mathematician become a pioneer of business as healing? Raised with a strong caring conscience in the Jewish tradition of *tikkun olam* (to repair or heal the world), Bernie read Philip Kapleau's 1965 classic, *The Three Pillars of Zen,* and soon began studying with renowned Buddhist teachers including Taizan Maezumi-Roshi at the Zen Center of Los Angeles. Bernie translated Zen teachings into his own version of the classic Three Tenets:

1. *Not knowing:* Akin to the notion of innocence described in the chapter on Jaipur Rugs, often expressed in the Zen tradition as "beginner's mind."
2. *Bearing witness:* Being fully present, without denial, rationalization, or opinion, in the midst of suffering.
3. *Taking action:* For Bernie, creative, loving action emerged naturally from the practice of the first two tenets.

On a visit back to New York, he was deeply moved by his encounters with homeless people living on the streets in the bitter cold of the Northeast winter. His beginner's mind and practice of bearing witness initially gave rise to a form of action that he called Street Retreats: bringing his students to live with and help care for the homeless, one week at a time. Recalling his experience living with the homeless, Bernie said, "The people we were serving were my teachers."

Subsequently, in the mid-1990s, Bernie started regular Bearing Witness Retreats at the sites of the Auschwitz and Birkenau concentration camps. He brought together people from many different countries, cultures, and religions, including Holocaust survivors and their children and grandchildren, as well as the children and grandchildren of the SS guards who had run the death camps.

Bernie understood, embodied, and shared the truth that human conscience is awakened when we bear witness to suffering and that this naturally gives rise to healing actions. He explains, "When we bear witness, when we *become* the situation—homelessness, poverty, illness, violence, death—the right action arises by itself. We don't have to worry about what to do. We don't have to figure out solutions ahead of time. Peacemaking is the functioning of bearing witness. Once we listen with our entire body and mind, loving action arises."[1]

In 1982, Bernie's loving action became focused on the community of Yonkers, New York, which at that time had the highest per capita homelessness rate in the country despite being located in one of the wealthiest counties in the nation.

Determined to create a solution to the connected problems of homelessness, addiction, and incarceration that was scalable and sustainable, he looked at the issue holistically. It was obvious that he needed to create housing for the homeless, but he also knew that unless the people who lived there had jobs, those houses would soon turn into slums, as had happened with most public housing projects around the country. But to be able to hold down a job, the large number of single parents that were part of the homeless community needed reliable childcare. Many of them suffered from drug addictions, so there was a need for counseling and family therapy. This was also the time of the onset of the AIDS epidemic. So Bernie decided he would also create housing for people with AIDS as well as provide health services for them.

Bernie wanted to do all of this in a self-sustaining way, not as a charity. He needed to create a business that involved relatively unskilled work that could be done by people hired from the streets. Inspired by the success of a Zen-influenced bakery in Northern California, Bernie decided to create a bakery. He initially started Greyston Bakery in the Riverdale mansion where he taught the Zen community. He decided to leave after neighbors complained when he brought in poor people to work at the bakery because business was greater than his students could handle.

After selling the property, Bernie moved to Yonkers to be immersed in the neighborhood he wanted to serve, buying a dilapidated industrial structure that had once housed a lasagna factory, as well as a nearby house. Thirty of his Zen community members moved to Yonkers, but many were reticent. As Bernie recalls, "The block that the bakery was on was loaded with crack vials. Next door to us was an all-night brothel."

He developed a comprehensive model that he called the Greyston Mandala, which would combine a for-profit business with multiple nonprofits that would work together to provide all the needed support and services. The centerpiece was the bakery, which would provide jobs, job training, and profits that could be invested in the welfare of the community. The second key piece was the Greyston Family Inn, responsible for housing and support services. Today, profits from the bakery support an interdependent system that provides jobs for 130 people, affordable housing for 530 residents (35 percent of whom were formerly homeless), childcare for 130 children, and housing and health care for 50 people with HIV/AIDS. They also operate six successful community gardens.

Greyston Bakery pioneered the idea of "open hiring." Their motto is: *We don't hire people to bake brownies, we bake brownies to hire people.* Anybody who wants a job simply adds their name to a list. The company hires and trains the next person on that list, without any background checks or even an interview. It hires people based on good faith, trusting that the vast majority of people, when given a job, will do it and that their skills and salaries will grow over time.

New hires at Greyston start as apprentices; they go through a ten-month job training and life skills course. The company works with each employee to develop a career path and then supports them with the necessary training, including GEDs and financial literacy, so that they can advance beyond entry-level jobs, or move on to well-paying jobs at other companies, which opens up space for more apprentices.

Open hiring is a life-changing healing policy that addresses one of the great tragedies of American society: the extraordinary number of people who are arrested and incarcerated. Approximately a third of working-age adults—over 73 million people—are listed in the interstate identification index, which means they have been arrested and fingerprinted by a local, state, or federal law enforcement agency.[2] That is roughly the same as the number of Americans with a four-year college degree. Nearly half of black males and 40 percent of white males have been arrested at least once by the age of twenty-three.

For most of those 70+ million individuals, the vast majority of jobs are off-limits; approximately 88 percent of employers refuse to accept applications from anyone with a criminal record. Even someone who was convicted at a young age of a nonviolent offense, served his or her time, and is now seeking to reenter society will find most doors closed. An estimated 75 percent of formerly incarcerated people find it almost impossible to get a job. Without constructive options, many return to crime, are re-arrested, convicted, and returned to jail at an average direct cost to taxpayers of more than $35,000 per year.

The tragedy is exacerbated by the lack of opportunities for those who grow up in high-crime, drug-infested communities. Bernie was inspired to hire the unhirable because, as he explains: "I want to give people a first chance. So many people in our country never even get that."

Greyston is not handicapped by hiring workers with such difficult backgrounds. By providing opportunity, hope, training, and respect, Greyston generates employee engagement that would be the envy of any *Fortune* 500 company. Employee turnover is only 12 percent, compared to 30 to 70 percent in similar manufacturing and production industries. Greyston saves approximately $2,500 per hire because it avoids recruiting and onboarding costs and does not do background checks and drug tests. This money is instead used for training and support and to pay better wages. Within a few years, apprentices who start at minimum wage earn salaries of approximately $65,000 a year with full benefits.

Greyston employee Dion Drew explains the difference that Greyston has made to his family:

I grew up in the projects in Yonkers, New York. At a very young age, I saw a lot of drugs being sold and used every day. My mom worked very hard. She went to work every day. But I guess the pay wasn't enough, because we was in a struggle, and as I got older, we continued to struggle. Meantime people in the streets seemed to be winning so I started selling drugs at the age of fifteen. I have been in and out of jail since the age of seventeen. My last term was four years, from 2004 until 2008.

When I came home, I looked for work every day, but nobody would hire me. As months went by, I started getting frustrated and angry. I was feeling hurt. A friend told me about Greyston Bakery, so I went down and put my name on the list.

I remember the day like it was yesterday. I was riding around with a friend, and I was ready to start selling drugs again, when I got a call from Greyston Bakery asking me if I wanted to work. I said, "Absolutely!" I been there ever since.

I'm very proud of that. To me, it ain't all about the money. It's about being a man and doing the right thing. And that's what I'm doing. I'm grateful for that. I have achieved everything I set out to accomplish. I have a beautiful three-year-old daughter, and my mom calls me at least twice a week to let me know how proud she is of me. I can't really explain the things Greyston has done for me. It saved my life. If I was still in the streets, I would be dead or in jail.

Dion is now a training supervisor.

The Greyston Mandala generates many similar stories. Gary White was released from prison in 2014 after being incarcerated for a decade. He saw a flyer for Greyston at his drug treatment program and put his name on the list. As he says, "If I had known about Greyston before, I never would have gone to prison."

Shawna, a single mother, was trying to find night-shift work to help support her five children, including her oldest who is afflicted with cerebral palsy and deafness. She explains,

At every interview, I was told that I needed experience or specific training. I tried to make ends meet by doing hair and babysitting but just could not keep up. I just was not able to find a job. . . .

I became depressed and overwhelmed. I felt trapped with no options but to consider giving my children up. . . . This is when I found Greyston.

Shawna adds,

Open Hiring sounded too good to be true after all the rejection I had faced. . . . Greyston saved my entire family. I have never been as grateful to anyone as I am to the people at Greyston. I feel like a new person: good about myself and my future.

Greyston CEO Mike Brady is a passionate advocate for open hiring and gets emotional when he recounts the stories. "Some employees were sleeping in the park and didn't tell us. When we found out, we helped them find housing. It is not just a job we're concerned with. It is the person. They should have the chance to succeed in life. There is a difference between equality and equity; not everyone needs the same thing. It is not only about Dion. It's about his daughter. He could go back to the street and make more money, but doesn't do so because of his daughter. She now has a better chance of going to college someday."

Greyston has now launched the Center for Open Hiring in Yonkers, and is launching a movement to get more companies to implement similar practices. Nearly fifteen organizations and businesses have joined the center, including Unilever and NYU's Stern School of Business. Ben & Jerry's, which has long been a customer, is considering implementing open hiring. It has already removed the checkbox on employment application forms that asks about the applicant's criminal record.

Bernie Glassman died on November 4, 2018, but he continues to inspire a remarkable range of humanity. If you visit Greyston, you will meet newly graduated MBAs and people with lengthy prison records; experienced, professional bakers and people who have never previously touched flour. They are all part of an enterprise that generates a consistent profit, and they are all united by a common vision:

Use the power of business to heal, to redeem, to give hope, to enable people to do what they never thought possible. Go from being wards of the state to contributing members of society. Go from homelessness and despair to a life filled with hope and optimism for themselves and their children.

THE MOST INTERESTING WOMAN IN THE WORLD

HOW EILEEN FISHER EMPOWERS WOMEN
TO MOVE FREELY AND BE TRUE TO THEMSELVES

t's not how you feel. It's how you look. And, Dahling, you look Mahvelous!

Billy Crystal delivered this line in a classic 1985 *Saturday Night Live* sketch entitled "Fernando's Hideaway," inspired by Fernando Lamas, an iconic playboy and movie star who appeared years before on *The Tonight Show* uttering the words: "It is better to look good than to feel good." Lamas was also the inspiration for his friend, actor Jonathan Goldsmith, to create another, more contemporary, charismatic playboy caricature: The Most Interesting Man in the World.

Crystal satirized a sentiment that runs deep in Hollywood and elsewhere: that "looks"—one's superficial appearance—is more important than feelings—one's inner experience. This notion is particularly oppressive and hurtful to women and is foisted upon them, as Jean Kilbourne and others have documented, by businesses who aggressively promote unrealistic standards of feminine beauty to sell magazines, food, alcohol, cars, plastic surgery, and, of course, fashion.

Eileen Fisher, one of the most interesting women in the world, has made a major contribution to healing the ensuing pandemic of body dysmorphic disorder, anxiety, and normative discontent. In 1984, with just $350 and four simple patterns, she launched a women's clothing business that now employs more than 1,200 people, with 56 stores and annual revenue of close to half a billion dollars. Fisher's success is founded on a simple but profound insight that remains in the sweet spot of the zeitgeist: An increasing number of women do not want to be identified, or form their identities, based *primarily* on how they look. Rather, they want to feel comfortable with who they are as multidimensional human beings.

As Eileen expresses it:

> The women who wear our clothes want to experience the magic that happens when you put on a piece of clothing that has been pared to its simple, pure essence. It comes alive on your body. It makes you move differently. It changes the way you think and feel about yourself.

In our research for this book, we interviewed many extraordinary individuals, but when we happened to mention to various female friends that we were speaking to Eileen, we were amazed at the fervor that ensued. We discovered that many of our female friends are not only loyal customers who build their wardrobes around Eileen's designs, they're also inspired by who she is and what she represents for them.

A physician friend enthused:

> Eileen's clothes make me feel professionally well dressed but still able to dance, laugh, imagine, and move in endlessly creative ways. There is a real understanding of what women actually do in a day and how to create clothing that effortlessly supports and empowers rather than a wardrobe that demands time, fuss, and body restriction. I also appreciate the corporate emphasis on sustainability and integrity as far as sourcing and production.

Another friend, a psychotherapist and mother of two daughters, expressed it this way:

I was walking through Grand Central Station the other day and happened to see the giant advertising posters for Eileen Fisher. I love seeing the images of healthy, positive, strong, and happy-looking women. This clothing honors positive feminine energy—something the world needs desperately. I want my daughters to grow up with this kind of inner confidence and healthy self-image.

And another friend who operates a yoga studio enthused:

OMG! I LOVE her clothes. Classic lines with just the right amount of quirk. Surprising combinations of fit and flow. And I love what she does for women and girls.

Eileen Fisher didn't start out with a desire to champion the archetype of the positive feminine by creating a Healing Organization that empowers women. She just wanted to feel more comfortable herself, having grown up in a home where she often felt awkward and constrained.

Raised in a Catholic family, Eileen was one of seven children. Eileen reflects that her mother never really had the opportunity to define herself. Her identity was obscured by the burden of child-rearing, which overwhelmed her. Feeling lost and depressed, her mother blamed her father for her unhappiness. Eileen was sensitive to her mother's plight and the discord between her parents. She grew up feeling a deep sense of discomfort associated, she realized much later, with a suffocating sense of externally imposed limitations.

Dance provided a refuge. Although she never took formal dance class and protests that she has never been a "good dancer," she nevertheless always loved to dance, to move freely. She explains, "As a child I experienced a sense of freedom and pure joy from just dancing around."

For Eileen's mom, sewing was a refuge. Eileen explains that, although her mom was often unhappy, "When she sewed, or went to fabric stores and we would touch the fabric, there was this aliveness and joy in her. It was a little place we connected."

Eileen's mother made her hand-sewn shift dresses that made it easy to move, to dance. And the seed was sown for the connection between clothing, freedom, love, self-expression, happiness—and, ultimately, healing.

How did she leverage these early experiences to create a company whose products are so healing to her customers?

Originally trained as an interior designer, Eileen understands the relationship between the design principles of form and function, and the human experience of feeling. She reflects,

From the very beginning I was quite an uncomfortable person. I needed comfortable clothes because that gave me a little more freedom and a little more ease in life. The ideas of freedom, ease, comfort, fluidity—all feminine ideals—are all embedded in the clothes. My emphasis was always on the feeling, the experience of actually wearing the clothes: the fabric, and how you feel in your clothing and allowing yourself to really feel what you feel, as opposed to being an object.

Eileen confides, "I guess I started designing the clothes really to heal myself."

Although she had elevated ideals and was building a successful business, her path wasn't easy. Eileen reflects that she unconsciously repeated some of the unresolved patterns from her parents' relationship in her own life. Feeling overwhelmed in her own marriage, she blamed her husband, who also worked at the company, for all the things that were going wrong. When they got divorced, she felt that her life was falling apart, and realized that blaming her now ex-husband wasn't the path to happiness or empowerment in the next phase of her life.

Eileen began to take personal responsibility, to look inward.

Soon after the divorce, she went to the Omega Institute in Rhinebeck, New York, to relax, read, and reflect. Elizabeth Lesser, one of the founders of Omega, knocked on her door and suggested that Eileen might get even more out of her retreat if she took a workshop.

It just so happened that one of the available workshops that week focused on dance. Although she had loved to dance as a young girl, somehow over the decades she had stopped dancing and become increasingly uncomfortable and disconnected from her own body. After five days of dancing, Eileen felt reborn! After decades of being in her head, she embarked on a journey of self-healing through embodiment that continues to this day. She recalls,

> In addition to dance, I started going on yoga and meditation retreats. I adopted practices that would help me to ground and get anchored and to heal myself, including psychotherapy. I don't think the business would've turned out so well if I hadn't been in therapy, because I don't think I would've been able to focus.

Of course, many people feel uncomfortable when they are children; and everyone projects unresolved childhood issues onto their partner when they get married. Patterns of dysfunctionality abound, but often remain unconscious, driving much conflict and suffering in personal and professional life. Some people become awakened and commit to growth and change through self-awareness, rather than acting out or giving in to addictions. Artists transform the experience of suffering into something creative, many therapists engage in their profession as a means of dealing with their own inner struggles, and leaders of Healing Organizations are involved in creating businesses that provide solutions for human needs in sustainable, profitable ways.

With her husband no longer at the company, Eileen began expressing herself fully through her designs and her vision of what the business could be. She developed her artistic expression while nurturing a therapeutic working environment, combined with a powerful business vision.

She proclaims, "The more I get freedom, the more I want that for the people around me."

Healing oneself is a never-ending journey. Eileen continues to work on her self-awareness, personal growth, leadership, and communication skills. She journals, meditates, and does yoga every day and has created an unusual corporate culture where self-development is encouraged and supported. For Eileen, this is a way of life but it's also good business. She explains,

> Projects fail when people don't work well together. So when you heal relationships, then you free up the energy and amazing work happens. We offer our people support from skilled coaches and we do workshops on giving and receiving feedback. We do a lot of work around helping people to get along, work together, and understand each other.

Eileen empathizes with the struggles of the younger employees in her company, and is passionate about helping them grow in confidence and self-determination. She loves to inspire them to become more effective leaders, by helping them cultivate the ability to interact in more constructive, joyful ways with their colleagues.

The company does a lot to help people understand themselves and discover their own purpose. In addition to communication and feedback training, values clarification and self-awareness workshops are offered to all employees, not just leaders. This gives people a common language in terms of understanding themselves, others, and their relationships. The company retains a consultant who offers a special program to help employees understand how their unresolved, unconscious childhood patterns drive their behavior, something that Eileen realized early on that provided tremendous liberation for her creativity. Having healed many of her own childhood wounds, she aims at nothing less than helping all her employees do the same so they can emerge into their fuller selves and work together seamlessly.

Healing Organizations don't separate work and life, or the professional persona from the authentic human being; rather they strive to encourage the wholeness and fulfillment of all team members. Eileen prioritizes the well-being of all her people and their families. She explains,

We need to start with people's lives as they are because a lot of people are struggling or suffering. They are in pain from various factors. Some of these are from work, and others may be home-related: an illness, a child with a learning disorder, or perhaps financial difficulty. If we don't look for these kinds of issues, we will not see them, and if we don't see them, we will not do anything about it and therefore people will continue to suffer. And of course that impacts their ability to work.

Eileen Fisher has created a culture where personal challenges are confronted with compassion and care. As the mother of a child on the autism spectrum, Eileen has a special sensitivity to families with special needs.

Eileen aspires to model the behavior and priorities that she wants the company to reflect. When her assistant's son tried to reach his mom in the midst of an important meeting, Eileen insisted that she take the call. Having had her own struggles with her two children, Eileen was frequently pulled away from meetings over the years to deal with challenging personal situations. She has fostered a culture where everyone knows that they can put their children and their families ahead of work demands. Rather than detracting from effectiveness, this policy generates a passionate commitment to excellence because employees feel gratitude and respect for the way they and their families are honored.

In addition to promoting emotional and spiritual well-being, the company offers significant support for physical well-being as well. Since the health-care system doesn't provide people with the kinds of services the company would like them to have, everyone is given an extra thousand dollars a year for wellness treatments and another thousand dollars a year for wellness education. They can use these funds for complementary medicine, acupuncture, massages, or energy healing. The company also brings practitioners into the office. There is a massage room and frequent on-site yoga classes.

The company is highly profitable, and approximately 40 percent of annual profits are distributed to the 84 percent female workforce in bonuses. The abundance also flows out to the world in wonderful ways: In 2017, the company gave $1.5 million in donations and grants to more than three hundred organizations whose focus is the empowerment of women and girls, supporting the fair treatment of all people, or preserving the environment.

Just around the time that Billy Crystal made "Fernando's Hideaway" a meme, Eileen's openness and curiosity led her to ask questions about more comfortable ways for women to dress, and ultimately more humane, environmentally sound, and intelligent ways to run a business.

She realized that how a woman feels in her clothes is of primary importance, and that a woman who feels good in what she is wearing will radiate an inner beauty. Eileen's epiphany when she renounced her Catholic school uniform back in Des Moines led her to create what she calls "a system of dressing that feels effortless and wonderful." You might even say: *Mahvelous!*

THE PARABLE
OF THE POTHOLE

HOW APPLETREE ANSWERS HEALED
THEIR DIVIDE BETWEEN RICH AND POOR

In 1995, John Ratliff founded Appletree Answers, a traditional telephone answering service. Operating out of his two-bedroom apartment in Wilmington, Delaware, Ratliff expanded the business through a series of acquisitions to more than twenty locations in the US. He employed well over six hundred people, 85 percent of whom were hourly workers and 15 percent salaried executives.

He shares the moment that it became clear that something wasn't right in the way that his hourly employees were treated in comparison to his salaried employees. On a chilly late-autumn day, one of the 15 percent, Appletree's CFO, was driving to the office when she hit a pothole and blew out her front right tire and damaged the rim. Fortunately, she was a member of AAA, and they dispatched a tow truck that brought her to her Mercedes dealership nearby. A "courtesy car" drove her to work and she had insurance to cover the cost of the repair. She arrived about ninety minutes late and was met by staff who offered sympathy and got her coffee.

On another day, one of the 85 percent, a customer service representative, was driving to work and hit the same pothole. She also blew out her front right tire and damaged the rim. Not being a member of AAA, she had to call a tow company and then a friend to drive her to work, since she couldn't afford a cab. She arrived at work nearly three hours late and was given a reprimand by her supervisor, who docked her pay and told her that she would lose her job if she was late again. Moreover, the repair for her damaged rim and new tire would cost more than she brought home in four full eight-hour days. With $160 in the bank, what was she to do? How would she even get to work the next morning? How would she drop her kids at school?

Ratliff recalls:

We were so empathetic to our CFO, and we rained all over the other person for whom this was a huge deal. We took her bad day and made it a lot worse. After the ordeal she was told: "Now go sit in your seat with a smile on your face and make sure that our customers have a great experience." It is totally backward. There was no empathy at all for the person who needed it most.

In 2008, an industry colleague asked him what his turnover was for hourly employees. When he replied that it was 118 percent, the colleague said, "Wow! That's fantastic. You are way below the industry average of 150 percent."

Upon reflection, John felt that his turnover rate for hourly employees was not an appropriate source of pride, even though it was better than the industry standard. When he discovered shortly thereafter that the turnover for his salaried employees was just 3 percent, he was shocked by the glaring discrepancy. John realized that he was presiding over two different working environments within the same company: one that people never wanted to leave, and the other in which people couldn't wait to escape. He reflects: "You can imagine the kind of class warfare this caste system created. If you really want to have a horrible culture, have your salaried employees be happy and your hourly people be miserable. Put them together and watch the discord play out."

After careful consideration, Ratliff and his senior team decided that they had to do a better job of caring for their hourly workers. Their work was challenging, dealing with one demanding, unhappy customer after another. Although they understood that unhappy people would impair the functioning and productivity of the enterprise and generate high turnover, they began to view the problem as more than just a tactical business issue. As Ratliff comments, "I began to realize how much suffering many of our people were experiencing at work. It was a real awakening of conscience."

Ratliff and his leadership group realized that the hourly employees who were leaving were from the lower side of the socioeconomic scale and the salaried people were from a higher strata. They saw that their company was a mirror for the imbalances in society and they became determined to find a creative approach to developing a culture that generated more engagement, wholeness, and fulfillment for everyone.

First, they got really serious about redefining and living their core values. Although they had done a lot of work to identify the core values and thought that they had made them part of the organization, by putting posters on the walls, cards on the desks, etc., they realized that the values remained abstract and weren't really informing the culture.

Ratliff, his COO, and the director of employee engagement decided to travel to the firm's different locations to meet all of the employees of the company. They announced a series of town hall meetings to communicate the values and generate engagement. Ratliff comments, "When you have 3 percent turnover in salaried positions, you are all very engaged and aligned and you think you are making great decisions. We celebrated our core value rollout strategy and we were all very proud of ourselves."

A few weeks later, when they got to the first office in Reston, Virginia, Ratliff offered twenty dollars to anybody who could recite all six of the values. After a long pause, one hand went up. She only knew two out of the six. Meanwhile, the values were listed on a poster right behind where they were sitting!

As they went from place to place, the same story repeated itself; people were getting only one or two of the core values right. Ratliff and his senior crew were becoming exhausted and demoralized, and he recalls thinking, "You would think that somebody would've called ahead and let other offices know that this question was coming!"

The last town hall meeting was scheduled at their office in Puerto Rico. On the way to the airport, John stopped at an ATM, and that particular machine was only dispensing hundred-dollar bills. This proved to be a significant quirk of fate. At the meeting, John asked his usual question about the values and a young man shot up his hand and waved it enthusiastically. He recited all six core values without any hesitation. John reached into his pocket and pulled out a hundred-dollar bill, which is all he had, and handed it over. The recipient started to cry from happiness. Everyone stood up and cheered.

That story spread quickly through the company. It was the turning point for their organizational culture. After that, the core values really started to come to life. Ratliff and his team made the values a third of the weighting of the score on employee reviews for everyone in the company, at all levels. Every time they gave feedback, whether for improvement purposes or to praise, they centered it around the values.

Ratliff comments, "Our employees realized that we were committed to living the values. They stopped being just a poster on the wall, and became integrated into the culture of the organization."

In addition to emphasizing the core values, Ratliff's team initiated complementary programs that helped to transform the culture. Most notably they created a program called Dream On—essentially a Make-a-Wish-type program for their employees. The idea was that anyone could submit a dream and that the company leaders would try to make it come true. They asked, "Please tell us the story behind your dream and why this is so meaningful to you." People could also ask for help to deal with life's potholes.

When they started the Dream On program, the executives who conceived it were patting each other on the back for being geniuses. But there was no response for the first ten days, probably because, as Ratliff speculates, people just didn't believe that it was for real.

After promoting the program internally, Ratliff's team finally received their first submission from one of their employees in St. Louis. She wrote, "I'm a single mom. My ex-husband stopped paying alimony two weeks ago. I got evicted from my apartment a week ago, and my two kids and I are living in my car. I'm so sorry but I need help. I just can't see my kids like this anymore."

Upon reading this, John's eyes welled up. He says: "I felt a deep sense of shame. I asked myself, 'What kind of company am I running here? A mother who works here full-time is homeless, and it took her this long to tell us?'"

Ratliff and his team arranged a hotel for her and her children on the same day they received her note and within two days they helped her get into a new apartment—normally, a huge hurdle for an employee without the money in the bank for a security deposit. He knew that over 40 percent of American workers have less than $400 in the bank; how are they supposed to come up with three months' rent?[1]

When Ratliff told her that the Dream On was confidential and she didn't have to share it, her response was, "Not going to share it? I want to tell everybody! No one has ever done anything this nice for me!" She shared her story, and it went viral companywide. Soon another employee under similar circumstances asked for help, and two days later she was in an apartment. Within two weeks, they had about fifty submissions. As Ratliff observes, "That was the domino that changed everything."

As the Dream On program gained momentum, they learned how to make it most effective. Every request was accepted for consideration. Some were granted the next day and others were granted months later, but all were acknowledged.

Over a four-year period, they fulfilled more than 250 dreams. Ratliff comments, "We weren't looking at this as an incentive program aimed at high-performers. Rather, it was based on a sense of caring and compassion for all of our people." He adds, "When you start to read some of these, you realize that so many things that we take for granted, other people are desperately struggling for—not just hourly employees, but salaried employees as well."

The effect on the business was dramatic. In a little under four years, their voluntary employee turnover went from 118 percent to 18 percent.

Ratliff comments, "It had so many positive consequences. Even if turnover didn't change, it still would have been incredibly valuable. It created so much empathy and positive energy inside the business." He adds, "We would get some really humble submissions from our hourly employees. For example, 'My dream is to have reliable transportation to get to work.' Really simple, but serious stuff."

Ratliff confesses that as an entrepreneur he was initially so focused on customers that he didn't understand or appreciate the people who were making his dream come true. The Dream On initiative changed that radically.

As more lives were changed for the better, the goodwill generated became palpable. As John explains, "Over time, a culture of empathy started to spread throughout the organization." As part of this initiative, they renamed their HR function the Employee Experience Department, whose primary job was to ensure that employees have a great experience at work, not just to be about compliance and risk-management. They hired a director of employee engagement, and a director of employee experience. This led to another healing practice: Management began asking every salaried employee who had direct reports to ask each of them one question every day: "*What action can I take now to make your experience as an employee better today than it was yesterday?*" This question became the organizational mantra. Ratliff exults, "Everyone started thinking all the time about what they could do to make the employee experience better for everyone. Anything that was an irritant, we tried to eliminate. It was little things and big things. It changed every part of our company, from the way we ran IT to payroll."

Many companies promote the idea of "open communication," but the mantra brought it to life. Call center work is challenging, but as employees discovered that their requests and needs were being addressed as a priority, morale skyrocketed.

Imagine that for years you've been sitting in an uncomfortable chair, your back hurts, but no one is interested; then, one day, your boss asks what you need to improve your experience at work. You say you need a new chair and then, by the end of the week, you have one!

Ratliff emphasizes, "Most of this is so simple. Yet, the power of this goes so far beyond what most of us comprehend. It is a fundamental responsibility of the entrepreneur in a business to create a place that adds value to a person's life and well-being, as opposed to a place that detracts value. In other words, it heals instead of causing disease. I have spoken about this to many audiences, and people come up to me and tell me that they have also tried it. Nobody has ever told me that it doesn't work."

In June 2012, he sold the company at a premium valuation, and now consults on how other companies can create healing cultures while generating high profits.

Many pundits pontificate on the "gap between the rich and the poor," with conservatives often arguing that everything is fine as it is and that poor people just need to work harder, while liberals frequently advocate higher taxes and redistribution of wealth. Both of these approaches fail.

Appletree didn't "redistribute wealth" or create a socialist system; rather, they extended caring, kindness, and healing with the result of more prosperity and happiness for all stakeholders. For most hourly employees, "working hard" isn't the issue; many hold multiple jobs to make enough to get by, but they're often victimized by predatory lenders and penalized by "fintaxes" that wealthier citizens have the luxury to avoid. Our next story explores how another caring entrepreneur is helping to heal this wound.

MAKING ROOM FOR DREAMS

HOW PAYACTIV IS HELPING THE WORKING POOR MOVE TOWARD THE MIDDLE CLASS

n 1998, journalist and activist Barbara Ehrenreich went undercover to learn firsthand about the world of minimum-wage workers in the United States. Accepting whatever jobs she could get, Ehrenreich moved from Maine to Minnesota to Florida, working as a nursing-home aide, waitress, chambermaid, and sales clerk at Walmart.

She sought housing based on what she could afford on her wages, which meant that she was limited to dreary residential motels and dilapidated trailer parks. She was shocked to discover how hard the work was. Ehrenreich, who holds a PhD in biology, explains, "the jobs were all mentally as well as physically challenging—I struggled to learn the computer ordering systems in restaurants, to memorize the names and dietary restrictions of thirty Alzheimer's patients, and, at Walmart, to memorize the exact locations of all the items in ladies' wear—which would then be rotated every few days." She also found that one job wasn't enough to cover even the most basic expenses and that like most of the working poor she would need two or even three jobs to get by.

In 2001 she released *Nickel and Dimed: On (Not) Getting By in America*, a book that became a major bestseller. For prosperous Americans, the economy may seem strong as we write this in early 2019, but the majority of people are financially weak, vulnerable, and living on the edge of disaster. Our sophisticated, technologically advanced, rapidly growing financial system systematically aggravates the burden on those who can least afford it by creating intermediaries that offer financing to meet short-term needs, to pay bills or buy a car etc., but do so in a way that almost guarantees the eventual financial ruin of their customers. The misery inflicted by usurious rates of interest are compounded by penalties and fees that suck the life out of those who can least afford them.

Financial Stress Data

- About 95 million Americans—78 percent of full-time employees—live paycheck to paycheck.

- Fifty percent have less than $400 in the bank, and would not be able to raise $2,000 within thirty days in the event of an emergency.

- Sixty percent are technically insolvent, which means that their liabilities exceed their assets. Most are going further into debt every year.

- Seventy-one percent of employees report that they suffer from financial stress.

HOW THE WORKING POOR ARE MUGGED

- *Overdraft fees:* These include returned check fees of $25 to $40 and overdraft protection fees. It adds up to over $25 billion a year, which is about 50 percent larger than the entire US music industry.
- *Bank fees:* Most working families in America can't afford to maintain a consistent minimum balance and therefore must pay $10 to $25 a month for a checking account. They pay each time they take out cash, and they pay to use a debit card. This adds up to $10 billion a year.
- *Pawnshops:* There are over 11,000 pawnbrokers in the US, generating nearly $15 billion in annual revenues with interest rates ranging from 5 percent to 25 percent per *month*, in addition to service charges that can reach 20 percent per month. A $400 loan could potentially end up costing nearly $1,000 a year in interest and fees.
- *Subprime credit cards:* A subprime credit card with $300 credit limit can incur a $49 annual fee, a $99 account processing fee, and an $89 program participation fee, along with a monthly account maintenance fee of $10.
- *Subprime auto loans:* Working-class employees typically live far away from the affluent areas where they work and need a car to earn a paycheck. But the same car that would cost a cash buyer $16,000 could end up costing over $25,000 when financed over five years.
- *Payday loans:* There are over 21,000 retail locations for payday loans. Most people get paid hourly but only receive their paychecks every two weeks. But many bills come due earlier in the month, so huge numbers of people run out of cash before receiving their next paycheck. For a fee in the range of 10 percent, they can access a few hundred dollars. If

they're not able to pay back when they receive their next paycheck, as is very common, the loan rolls over, and the interest compounds. Delinquent loans then get sold to collection agencies, who aggressively hound people to pay the full amount and more.[1]

These are just a few of the elements of what has been called a "fin-tax" on the working poor. There's an obvious undesirable effect of all this on the quality of life for families subject to this mugging.

According to the Center for Financial Social Work, financial issues are the greatest stressors in people's lives. They are the leading cause of divorce, and contribute to violence, alcohol, and drug addiction, a wide range of stress-related diseases, and homelessness.

And it turns out that, although many companies profit from all this suffering, in the bigger picture fintaxes are bad for business.

THE BUSINESS
IMPLICATIONS OF FINANCIAL STRESS

Financial stress causes employees to be distracted, and leads to high levels of absenteeism as well as "presenteeism"—when employees are present in their body but absent mentally because of preoccupations. It is estimated that financially stressed employees spend about twenty hours a month at work dealing with related issues. Companies with employees under financial stress experience higher turnover and health- care costs. High turnover means higher recruiting, orientation, and training costs for replacements.

In an article published in *The Nation* entitled "Preying on the Poor," Ehrenreich explains that the working poor aren't appealing targets for street muggers because they don't have much cash in their wallets. But she does show how big businesses often mug those who can least afford it. She writes:

Lenders, including major credit companies as well as payday lenders, have taken over the traditional role of the street-corner loan shark, charging the poor insanely high rates of interest. When supplemented with late fees (themselves subject to interest), the resulting effective interest rate can be as high as 600 percent a year, which is perfectly legal in many states.[2]

In the tradition of journalists Studs Terkel, H. L. Mencken, and others, Ehrenreich is gifted at inspiring righteous anger, and awakening readers' consciences, but where can we find solutions to the nickel-and-diming of the working poor?

SAFWAN SHAH'S EPIPHANY

Safwan Shah grew up in Pakistan, where as a child he witnessed societal chaos, riots, and the imposition of curfews and martial law. His father was a rare character—an honest civil servant who refused to take bribes—and thus he was "transferred" to a different city every year or two. Despite this itinerant childhood, Safwan was an exceptional student and graduated college with a degree in electrical engineering.

Seeking a better life for himself and his family, he immigrated to the United States on January 1, 1989. He soon completed a master's degree followed by a PhD in aerospace engineering. In 1994, he found his way to Silicon Valley, initiating a series of entrepreneurial ventures with mixed results. In 1999, he founded Infonox, a company that served casinos, enabling patrons to access money easily and securely. He helped develop an automated cashier machine (ACM) equipped with facial biometric cameras, ID scanners, fingerprint sensors, and check imagers and scanners. These ACMs could verify a customer's identity, cash checks, receive money transfers, issue credit lines, redeem rewards—most of the things that a human cashier in the casino could do. In 2008, Infonox was sold to TSYS, where Safwan remained for a couple of years as president before "retiring." He earned a business degree at Stanford and played a lot of golf for a few years. Money was no longer an issue, but meaning was.

He began, with his wife's encouragement, to ponder the question of how to utilize his insights, expertise, and experience to serve a higher purpose. One day as he mused on this question at a coffee shop, he considered the circumstances of the hourly employees who were serving him. He recalls,

It was thirty years ago that I first observed someone ask an employer for a salary advance. I was an intern at a manufacturing facility during my college summer break and a production line worker made this simple request to the supervisor: "It is the 27th of the month. Can I get ten days' salary, and can you deduct it on the first when I get paid?" Fair enough, I thought. After all, this individual had already earned their money. Yet, the retort from the employer was not to help but to scold. "You better take control of your life and manage your finances better. I am not your bank and not a lender." I was taken aback. . . . I never forgot it . . . someone in need had asked for money they had already earned, and they had been rebuffed.

Safwan became passionate about addressing this injustice. His epiphany can be summed up in the idea that for the nearly 100 million workers who live paycheck to paycheck in the US, the *timing* of money matters almost as much as the quantity. He asks,

What do people come to work for? It's fine to say when you are a white-collar worker that you come to work for autonomy, mastery, and purpose. . . . But when you are working for $10 to $15 an hour, you come to work to make ends meet: to put food on the table, to survive.

He adds, "When you are making $15 an hour, you're treated as a commodity. There's no one sending you TED talks to get inspired."

Safwan asked some basic questions:

- Why should over $100 billion sit in the system every week, earned but not accessible?
- What if even 5 percent of these funds were made accessible to the employees who needed them the most? In fifty-two weeks, this amount would translate to $260 billion.
- Would it help to reduce predatory fees, overdraft risk, and predatory short-term lending products?
- Why couldn't we use existing technology to remove the obstacles that stand between the time income was earned and its access?

Thus was born PayActiv.[3] The idea was simple: Give people access to money they have already earned. People are able to access money from "the most vested and logical stakeholder in their financial lives: their employer." This transforms the vicious spiral of debt, misery, presentism, and high turnover cost into a virtuous spiral moving toward solvency, dignity, and productivity.

The testimonials from users are numerous and heartfelt. This one sums up the power of PayActiv:

I have three kids and my husband and I both work. We have paid overdrafts at least four times, and a late fee on rent six times, in the last year. PayActiv has helped us avoid these fees. It makes it possible for us to begin living our dream of saving up so we can become homeowners someday.

Safwan is on a mission to awaken the consciences of CEOs. He points out, "You don't need to be borrowing money from your poorest employees." The moral case is clear. The Bible admonishes: "Pay them their wages each day before sunset, because they are poor and are counting on it. Otherwise you will be guilty of sin" (Deuteronomy 24:15 NIV). In Dante's *Divine Comedy* those who gouge the poor with high interest rates and fees become permanent residents of the

Seventh Circle of Hell. Safwan also makes a compelling case that the energy and engagement unleashed when hourly workers are liberated from the stress of fintaxes will more than compensate for the investment in PayActiv's service.

At the end of 2018, Safwan wrote to PayActiv's board, advisors, and investors:

> We have the greatest job in the world. In 2018 we served 442,158 workers and moved $537,479,371 into their hands through timely access to their already earned wages. We put an additional $110 million in the pockets of lower income workers. Most estimates are that we added $200 in monthly purchasing power for a typical worker. A 7 percent raise.
>
> With your help we give a helping hand to workers who are drowning, bringing them above water so they can breathe and get their vigor and bearings back. Yes . . . we may have the most sizzling business idea, but all that pales in comparison to the fact that we make mothers, fathers, and children smile.

WHERE ARE THE CUSTOMERS' YACHTS?

HOW FOOLS, A SHAMAN, AND LARRY FINK ARE BEGINNING TO HEAL WALL STREET

I n 1940, former stockbroker Fred Schwed published *Where Are the Customers' Yachts?*[1] The title comes from a question a visitor to New York innocently asked after being shown an array of yachts owned by Wall Street executives. Of course, there weren't any customer yachts. Financial advice is usually much more profitable for those who sell it than it is for those who buy it.

The 2008–09 financial crisis was an egregious example of the many dysfunctions at the heart of the financial sector, wiping out over $11 trillion of wealth for Americans in a haze of confusion and deceit about "mortgage-backed securities," "credit default swaps," and other exotic and malodorous creations.

The financial sector generates disproportionate wealth for its denizens while too often adding little value to the economy or to people's lives. Onerous, disguised fees and rampant conflicts of interest make it one of the main drivers of unconscious capitalism.

The culture of Wall Street is one of extreme pressure and stress, well depicted in the movie *The Wolf of Wall Street*. It is often predatory toward customers and abusive toward employees, even while many of those employees are exorbitantly compensated.

There are notable exceptions to the ethos of greed and prevarication, harbingers of an emerging better way.

Court jesters were believed to be the only ones who could speak the truth to monarchs. Brothers Tom and David Gardner decided, twenty-five years ago, to name their investment advisory The Motley Fool, emphasizing their commitment to telling the truth, and to a positive, playful attitude. Growing up in a stable and loving family environment with a tradition of entrepreneurial success and Christian values, they were introduced to the art of investing at an early age with a simple philosophy: buy-and-hold great, conscious companies for the long run. It has led to a successful business, for them as well as for their employees and customers. As David told us, "There is a lot of happy-go-lucky American optimism baked into us as brothers and therefore into our company, our brand, and what we're trying to do in the world."

The purpose of The Motley Fool is to "make the world smarter, happier, and richer." David says, "The business of finance should be a beautiful one. We think people who have saved money are heroic, because they have had the discipline to generate their own financial sustainability. Wealth management, when done properly, is a wonderful profession because you are working with heroes to make them more capable of achieving more things on behalf of their family, community, and country."

With the support of Lee Burbage, The Motley Fool's chief people officer, Tom and David have created a workplace where employees, known as "Fools," feel valued and supported. David told us, "We spend a lot of time with our fellow Fools trying to ensure that they are doing what they love and what they are good at. If you're doing something you're passionate about and that you're good at, and you're doing it alongside people who you love for a mission you believe in, that is the playground we all want to be in."

Every one of the now-320-strong complement of Fools has access to a life coach, who checks in with them a few times a year and asks,

"Are you happy? Are you getting to do what you want to do to express your highest calling?" This desire to support people in their quest to fulfill their talents and dreams has always been a part of The Motley Fool's culture. Ten years ago, one of the Fools on the technology team was considering leaving to pursue his dream of being a personal trainer, but instead he became the company's first full-time wellness advisor. When people are supported in their wellness and when they love what they do, they don't want to miss a day, so the company has found no need to draft a policy for sick days and has a "take what you need" vacation policy.

The company also helps customers heal from previously damaging financial experiences, which most have had. They guide them to understand the power of investing to help shape the future. David states, "When we put money into a stock or a startup, we are making a choice to give them life and not something else life. We are truly in micro-ways every single day shaping the future. We urge customers to invest their money in a way that is truly consonant with the future they want to see. We say: Make your portfolio reflect your best vision for our future."

Lawrence Ford shares The Motley Fool's commitment to help clients invest in ways that will fulfill their best visions for the future. He became a financial advisor right out of college and soon became one of the top producers in his company. At thirty-six, he had two boats, two maids, two Mercedes, and a big house in Connecticut next to a golf course, with a swimming pool in the backyard. But although he felt that his success was predicated on his genuine caring for his clients, he was, nevertheless, becoming disillusioned with the toxic culture pervading the industry.

Seeking spiritual growth and deeper wisdom, Lawrence left the world of finance and engaged in a three-year intensive spiritual quest that included being initiated in a Tibetan shamanic lineage. On the verge of becoming a monk, his inner guidance led him to come back to the world of finance and be the change he wanted to see in the world.

He returned to Connecticut and started a new registered investment advisory firm with the intention of helping clients heal the anxiety and worry that they often have about money. He realized that the key to doing this was to align people's portfolios with their values so that their investments were expressions of a sense of purpose; and in the process, deploying capital in the direction of helping to heal the earth.

Conscious Capital Wealth Advisors HQ feels more like a spa than a financial services company. The thirteen practitioners in the office include specialists in managing life transitions, physical fitness, nutrition, massage, and life planning. There are in-house classes in yoga, tai chi, qigong, and meditation. All of these are for employees as well as for clients. Lawrence explains, "You sign up for financial services, but you get all of this because we're trying to serve the whole person. If we get you to your goals and you are not healthy to enjoy them, then what good is it?" This exceptional holistic approach to investing led the *Washington Post Magazine* to feature him on its cover as "The Shaman of Wall Street."[2]

The Fools and the Shaman are relatively small operations but they are harbingers of a movement to bring healing to financial services. This movement is gaining momentum, as evidenced by the recent letter Larry Fink, CEO of BlackRock (whose $6.44 trillion under management make it the world's largest asset manager), wrote to CEOs. Fink emphasized that business must take the lead in healing the world:

> Unnerved by fundamental economic changes and the failure of government to provide lasting solutions, society is increasingly looking to companies, both public and private, to address pressing social and economic issues. These issues range from protecting the environment to retirement to gender and racial inequality, among others.

Fink makes a compelling case that a clearly defined purpose is critical for companies who wish to thrive in the long term. He explains, "Purpose unifies management, employees, and communities. It drives ethical behavior and creates an essential check on actions that go against the best interests of stakeholders."

Fink knows that we are at the beginning of the greatest wealth transfer in history, as boomers shift an estimated $24 trillion to millennials. He cites a recent survey of millennials in which they responded to a question on what the primary purpose of business ought to be by agreeing overwhelmingly that "improving society" was a greater priority than "generating profit."

Fink explains, "As wealth shifts and investing preferences change, environmental, social, and governance issues will be increasingly material to corporate valuations." And he concludes, "At a time of great political and economic disruption, your leadership is indispensable."

This isn't just true for CEOs, it's true for us all.

THANKS FOR PUTTING POISON ON MY MICROSCOPE

HOW HILLMANN CONSULTING REMEDIATES TOXIC ENVIRONMENTS WITH INTEGRITY AND CARING

C hris Hillmann founded the environmental consulting firm that bears his name in 1985.[1] He explains his motivation for creating his own company and how it evolved into a Healing Organization:

> My first job after college was with a firm out of Staten Island. Back in the bad old days, asbestos abatement in New York was controlled by "solid waste contractors"—also known as the Mafia. With my newly minted bachelor's degree in business I was anointed as their "safety" guy. But it soon became obvious that they had no interest in safety or human welfare. They were always screwing with my equipment and changing my results. They were crooks.

Chris adds,

> The company was focused on remediating toxins, but it was a toxic working environment. I couldn't stand working there. Shortly after I left, my former boss was arrested and did time for bribing an EPA official.

Chris resigned and secured a job with an environmental consultancy firm that he hoped would be a better place to work. He became an asbestos assessor and an abatement project monitor and was certified to analyze air and bulk samples under a microscope.

But then, on a project in a New York City office building, he detected dangerously high levels of asbestos in an occupied area. The building manager offered him $5,000 to change the result. He refused and his boss removed him from the project the next day.

Chris notes, "I had a queasy feeling that after my boss switched me off the project he had accepted the bribe himself."

Chris's suspicion that his new company was also corrupt was confirmed after he was put in charge of monitoring asbestos levels at a Catholic school in Delaware. He discovered unsafe levels of asbestos debris ground into the carpeting, and even between the pages of the books in the library. Chris reported these finding to his boss but then overheard the boss talking to a representative of the school and telling him it was just dust and debris and not asbestos. "I couldn't believe it! How could anyone be willing to put others, especially schoolchildren, at risk for mesothelioma and many other deadly ailments, in order to make a few extra bucks?"

Chris went into another office and called the school authorities with the truth. He gave his notice, but during the two weeks before he left, he was asked to analyze lab samples. As he was leaning over to examine a sample under his microscope, he noticed, just in time, that dispersion fluid (a dangerous irritant) had been rubbed on the two oculars—presumably to hurt him!

Chris walked out and never looked back. He conceived the vision of an environmental consulting firm that would actually care about the environment and about all its stakeholders. He thought, "There must be a better way!" Chris says, "I knew we could leverage integrity, science, and business acumen to stand out in this field."

Realizing that he needed a partner with a science degree in order to get a license to operate a lab, Chris asked his brother Joe if he could help. Joe, who had a bachelor's degree in earth and atmospheric science, moved up from Virginia to join him in northern New Jersey.

They bought used equipment, including their first microscope, and worked night jobs to get by as they completed the technical certification processes necessary to start an environmental firm.

Joe was the science guy and Chris was sales and management; they both served as technicians on every project. Chris recounts, "I remember going on sales calls in fancy New York high-rises, riding the elevator for forty floors, and comparing my hand-me-down suit with the elegant apparel of the other riders, and wishing I looked older and more established."

Chris's sartorial limitations didn't impede the growth of his business. After billing $360,000 in their first year, $1.9 million in the second and $3.2 million in the third, Hillmann took off, securing many of the largest developers in New York City as clients and expanding their services and geography.

It wasn't always easy, however, and the company had to rally to get through three recessions. Chris recalls the toughest time when it seemed that they might have to declare bankruptcy because their bank was cutting off their line of credit in the low point of a recession when they needed it most. He chuckles as he explains, "We didn't declare bankruptcy because we couldn't afford a bankruptcy lawyer." Chris adds, "The attorney we consulted did give us a piece of golden advice: He said, 'Get a new bank!' We did and it all worked out."

Chris adds, "We also fired a number of clients and declined to work with others when we felt that they didn't do business in an ethical way."

For Chris and his brother it was self-evident that honesty and integrity were the bedrock of their enterprise. They also just naturally believed that kindness, caring, and generosity were intrinsically valuable ways of being and that these values would help them be successful. Chris says,

We often compete against companies that primarily use free-agent individual subcontractors so they don't have to pay benefits or provide education and training to their employees. We chose a different model. Our thought was that by taking good care of our team, they would take the best care of our clients. We always felt we could do well by doing good.

Through their "Hillmann Cares" program, the company takes care of its people in many ways. Just in 2018, the company:

- Paid the cost of a successful rehab program for an employee who had become addicted to prescription medicines.
- Gave a number of no-interest loans—and even financial gifts—to help employees with a range of challenging personal situations like unexpected flooding or wind damage or replacing a broken furnace.
- Continued to pay the salary of an employee who was diagnosed with a terminal brain disorder and making arrangements for the employee to transition onto a sustainable long-term disability program.
- Gave six weeks of paid bereavement time to an employee who experienced the loss of a close family member.

In addition to being a source of financial and emotional support for employees who are facing hardships, Hillmann creates a healing environment by offering a generous creative employee wellness initiative, a customized leadership development program, a policy of sharing 20 percent of profits with employees, and most importantly a sense of higher purpose. The company purpose statement is simple and clear:

Making a better future for all the communities we touch.

It's not surprising that Hillmann Consulting appears every year on many registers of "Best Places to Work." Like other Healing Organizations, Hillmann's focus on the welfare of its own people translates into service to the community.

From the beginning, Chris and his brother contributed a significant portion of their profits to worthy causes in their local community.

Over the years they've been consistent supporters of Habitat for Humanity, the Boys and Girls Clubs, and other charities, but as the company has expanded to include offices in Boston, Virginia, California, and other locations around the country, Hillmann's executive team now gives each regional office an annual charity grant for them to use on local causes.

And the spirit of giving manifests by more than just donating money. Hillmann employees contribute their time, energy, creativity, and expertise to everything from assembling and distributing "apartment baskets"—pots and pans, dishes, silverware, a vacuum, sheets, towels, blankets, cleaning items—for less fortunate families transitioning into affordable housing projects to donating construction advisory services to a women's shelter.

All of this goodness makes Hillmann a wonderful place to work; it gives the company a competitive edge both in terms of attracting and retaining the best people and also in generating long-term client relationships.

Chris comments:

We've always been committed to a culture of caring, but as we grew to multiple offices in different geographies and multiplied the number of folks in the company beyond the point where it's easy to get to know and interact with everyone individually, we realized that we needed to define our values and communicate them consistently and intelligently.

Chris and his sales and business development team discovered an unexpected benefit arising from the work put into articulating their purpose and values:

Now, when I'm on sales calls and I share with potential clients that we are in business to make our communities better, they start really paying attention. They get that we are really serious about this and start taking copious notes. At first this shocked the hell out of me because we are in a business that is traditionally very price-sensitive where people are often just looking for the low-cost provider . . . but the effect of communicating our passion for caring brings something out in clients that makes them want to do business with us and they're willing to pay us more!

• • •

Stephanie Cesario, Hillmann's managing director of Construction Services, led the internal effort to define and communicate the essence of Hillmann's culture of caring. She comments:

When you're dealing with human health and safety and construction quality you want people who really care. The low-cost providers in this space don't have caring committed teams; they have a one-off, short-term approach. Clients recognize that in the long term they will save money by working with us.

Matt Kamin, head of Hillmann's Construction Risk Management group, adds, In our annual strategic planning meeting three years ago, we identified 'recruiting and retaining the best people' as our greatest business priority. We also made a major commitment to become a more conscious, healing organization. This commitment was then expressed in a video we produced highlighting our culture of caring. Last week I showed it to a candidate who had offers from two of our competitors for equivalent compensation. But when he saw the video about everything we do to make the world a better place, he realized we are for real and he's now working for us. When we first began this journey to be a Healing Organization, I confess I was a bit skeptical but now I'm all in. We've had tremendous growth since we began this effort, 40 percent just last year, and last year was also by far our most profitable.

NOT "ONLY" FOR PROFIT

HOW KIND SNACKS MAKES "HEALTHY" DELICIOUS AND PROFITABLE

Roman Lubetzky, father of KIND founder Daniel Lubetzky, was just eleven years old when he and his family were imprisoned in a Nazi concentration camp. On the verge of death by starvation, Roman remembers an unexpected act of kindness that he believes saved his life: A guard at the camp threw him a rotting potato. That doesn't sound very appetizing but for the starving child it provided enough sustenance to keep him alive; moreover, he knew that the guard had risked imprisonment or even death by helping a Jewish prisoner.

After being liberated from Dachau at age fifteen, Roman took refuge with relatives in Mexico City. As he slowly regained his health, he began to learn Spanish and English and read voraciously. He became a successful entrepreneur, got married, and started a family.

Daniel was born in Mexico City in 1968. When he was sixteen, the family immigrated to San Antonio, Texas, partly due to increasing anti-Semitism in Mexico. After graduating from Trinity University in San Antonio with a degree in economics and international relations, Daniel studied abroad in France and Israel and then earned his JD from Stanford Law School in 1993.

Daniel enthuses about how his father inspired him:

He saw life as a gift, as a blessing, since a lot of people around him perished in the concentration camp. He felt he survived thanks to the kindness of others. His life mission was to be kind. If someone was having a bad day, and was moody—a flight attendant, a waiter—he would tell jokes unrelentingly until he got them to smile. He was here to give light to the world and he inspired me in many ways.

Daniel's father also encouraged his entrepreneurial spirit. Daniel started his first business at the age of eight, performing magic shows as "The Great HouDani." He worked more magic by starting a number of successful business ventures while he was still in college. Inspired by his father's example of kindness and conscience in response to grave suffering, he always felt that business could be more than just a means to make money; it could be a force to help heal the world's biggest problems. From the time he was around twelve years old, while he was preparing for his bar mitzvah, Daniel was obsessed with finding solutions to the Arab-Israeli conflict, and later wrote his college senior thesis on how the problem could be solved through conscious capitalism. He wrote, "The fundamental idea is to use business as a force for bringing people together, shattering stereotypes, and mending relations."

He put these ideas into action with a company called PeaceWorks, which he launched in the early 1990s after finishing law school. The idea was to get Palestinians, Israelis, Jordanians, Egyptians, and Turks working together with the common goal of producing various healthy delicacies that are enjoyed by Middle Eastern cultures, such as sun-dried tomato spreads and various pestos and tapenades, which at the time were relatively new to American palates.

Daniel learned many lessons through PeaceWorks, most notably that having a good cause isn't enough; you've got to create a product that people really love. As he expresses it: "You can have a product made by Mother Teresa, but people won't buy it again if it isn't the best."

PeaceWorks is still in business, and in 2004 Daniel launched KIND Healthy Snacks, a company whose delicious, nutritious, and healthy snacks now generate roughly $1 billion at retail. This "not-*only*-for-profit" business is at the center of a movement to spread kindness around the globe through a number of social initiatives.

Daniel explains, "It is clear that social objectives, if they are authentic, can actually enhance the brand and the value of the company. But much more important than that, what are we here for? Are we here just to make money? We're all going to die. We're not going to take that money with us. What can we do to make sure that the world for our children is better than how we found it?"

Daniel is devoted to the ideal of the Healing Organization. He is obsessed with helping others think creatively about how to design businesses that help achieve our most critical social objectives in a way that is sustainable and scalable. He explains, "The business benefit is not the only one, and increasingly in my way of thinking, not even the primary one, but is the result of building something that is really transcendent."

This transcendental approach is grounded in a solution-oriented, creative mindset that looks at issues from the perspective of *both/and* rather than *either/or*. Habitual thinking dictates that we can have *either* a tasty snack *or* a healthy one, either a convenient food *or* a wholesome one, *either* a profitable company *or* a socially conscious one. But KIND bars are truly delicious *and* good for you; they are convenient *and* wholesome, and the company is financially successful *and* is making the world a better place!

And, leaders can be kind, compassionate, caring and strong, focused and powerful. This integration of seeming opposites is fundamental to the KIND culture.

Being kind does not mean that you are soft and not committed to excellence. Kindness and excellence go hand in hand. Being kind requires tremendous strength and having the courage to stand up to injustice, and to lend a helping hand to others.

Being kind is not the same as being nice. Running a company based on being nice can get you into a lot of trouble. You don't share difficult feedback or take tough decisions that you need to because you're being nice and you end up being mediocre or going out of business. Kindness requires strength, courage, candor, and caring.

KIND's hiring practices are essential to maintaining and scaling these values in action. The company actively seeks candidates who manifest positive energy and caring, who have a sense of purpose, and who are fundamentally kind.

Positive, healing cultures don't come about just because someone drafts a great mission or value statement. The leadership of the company has to truly embody the noble sentiments and reinforce them continually. This is certainly the case at KIND: "We prize collaboration and cherish team spirit, so we continually nurture and invest in it. We never take it for granted, we talk about it constantly, and remind everyone all the time."

KIND employees, in Ken Blanchard's classic phrase, "catch each other doing something right." They send one another notes called *Kindos* (like kudos), and those who most embody the company's values are recognized as *Kindos of the Year.*

On the rare occasions when KIND needs to let an employee go, they don't use the term "firing." Daniel explains, "You will never see a harsh transition where somebody is given a cardboard box and escorted out of the building. Other companies do that routinely." He adds, "We've created a culture that just doesn't attract selfish people. There are no jerks here."

Daniel is seeking to find the dynamic balance between the rapid growth of KIND and the humanistic values he cherishes. He explains,

In our economic system, staying the same size is commercial blasphemy. You cannot decide that you have a really nice business and don't want to grow, because in the consumer product space, they measure you based on your shelf space performance. If you're not selling more than your competitors, they are going to discontinue your product. It is a very merciless world; either you are winning and taking market share from your competitors, or they are taking it from you.

He adds,

That is a very dangerous way to live, that everything is about who can be the best, who can grow the fastest, who can achieve the most efficiency, because it drives us to a consumer world where consumption ends up defining some people's happiness and we end up where we are today. I don't have the answers yet for how to tackle this, but it has to include a redefinition of how we define business success.

• • •

It would have been understandable if Roman Lubetzky became a bitter and vengeful person after losing much of his family and a good part of his childhood, and then experiencing anti-Semitism in the country to which he escaped. But instead he chose to affirm life, humor, and love and he passed these qualities on to his son.

In the depths of hell-on-earth, a small, unexpected act of kindness helped to save a boy's life. It's a long way from a rotting potato to a dark chocolate cherry cashew snack bar. But as the fabled Greek poet Aesop (620–564 BC) wrote,

The level of our success is limited only by our imagination and no act of kindness, however small, is ever wasted.

KIND is one of the fastest-growing brands in the snacks market. Mars bought a minority stake in the company in 2017, valuing it at over $4 billion.[1] More than a billion KIND bars were enjoyed in the last year and innumerable acts of kindness have been inspired by Daniel's initiatives. And these initiatives are part of a movement to reimagine the notion of business success so that it always includes kindness, healing, and love.

REDEFINING
SUCCESS

HOW *CONSCIOUS COMPANY* MAGAZINE
LIVED UP TO THEIR OWN VALUES AND PURPOSE

n March 2014, Meghan French Dunbar was enjoying a slice of pizza and a glass of red wine with her friend Maren Keely at Pizzeria Locale in Boulder, Colorado. A year earlier, she had received her MBA with a focus on sustainability in business and found a job working in the publishing industry. In the flow of their conversation, Maren, who was then studying for an MBA with a focus on how business could be a force for positive change, asked Meghan: "Why doesn't a magazine exist for conscious businesses? There's *Inc.*, *Entrepreneur*, *Fast Company*, and *Success*, but where's the magazine for businesses that are actually trying to do good in the world?"

Meghan's first thought was, "Surely, such a magazine exists. There must be one!" It seemed like such an obvious gap in the market. They returned to Meghan's house after dinner, pulled out their computers, started searching, and discovered, to their amazement, that nothing like that existed in the marketplace.

The two friends talked for hours about the possibilities and got incredibly excited about launching a magazine focused on business as a catalyst for a better world.

That night Meghan dreamed about the idea and was inspired, even euphoric, about the potential.

When she woke up the next morning and checked her email, she was shocked to discover that she had been fired from her job. It was completely unexpected. Meghan reflects,

> Instead of having crazy, chaotic, fear-based thoughts like, "Holy shit, I have been fired. What the hell am I going to do?" my immediate response was, "Yes! This is the universe telling me that I am supposed to start a magazine about sustainable business!"

It took a few days to convince Maren, whose skepticism about messages from the universe was grounded in her keen awareness that they had little experience or money. But Meghan's enthusiasm and sense of mission was irresistible and together they began to figure out how to proceed. First, they needed a name, and that came easily: *Conscious Company* magazine. The domain name was available and they set up their new email addresses. Then, they made a wish list of twenty conscious business leaders they admired and would want to feature in the magazine. At the top of the list was John Mackey, founder and CEO of Whole Foods Market and cofounder of Conscious Capitalism, Inc.

Meghan and Maren wrote to all twenty leaders, letting them know that they were going to press in January 2015, and asking them if they would participate. Much to their combined surprise, horror, and delight, nineteen of the twenty said yes, including John Mackey.

Suddenly, what had seemed like a long-shot dream for the future was now a tangible, compelling, urgent priority. Meghan reflects, "Up to that point it felt like we were playing chicken with each other about who was going to stop first. But, then, when John Mackey committed, things got real, fast. We both realized: Wow—we actually have something here!"

Determined to create a well-curated, high-quality print magazine, they realized they needed a production plan, money for a designer and logistics, and of course lots more money to actually print the magazines.

They launched a thirty-day crowdfunding campaign on Kickstarter, with the goal of raising $50,000. Money started flowing in from all over the world, pledged by people inspired by the vision. But the way Kickstarter works is that if you don't reach the goal that you set, you don't get *any* of the money. With a few days left, they had commitments totaling $42,638—and then the money stopped flowing in. They were $7,362 short, but were sure that this had to end the way it would in the movies, with somebody swooping in at the last moment to put them over the top. That didn't happen.

Meghan recalls, "We sat on the floor of Maren's apartment and watched our campaign go down the drain. In its beautiful infinite wisdom, Kickstarter actually starts counting down by the second for the last two minutes. It was like waiting for a bomb to go off . . . tick, tick, tick . . . and then flashing on our screen: 'I am sorry, you failed.' Just like that, our $42,638 vanished."

It was a crushing blow. Although friends and family were supportive and sympathetic, many were also secretly relieved that perhaps now these two promising young women were not going to squander time and money in a quixotic quest to launch a print magazine in the digital age.

Meghan and Maren took a week off to regroup and figure out what to do next. When they returned, they found a flood of emails from people all over the world who had been part of the Kickstarter campaign urging them to continue. The gist of the notes was: "You have to do this. The world needs you! Just launch a new crowdfunding campaign. Lower the goal. Let us know; we will throw the money back in." So they launched a new campaign, this time on Indiegogo, and raised $20,000. They hired their first employee (a designer) and miraculously got the first edition of the magazine published on schedule. With John Mackey on the cover, the premiere edition of the magazine was displayed in every Whole Foods Market in the country, and people loved it!

The good news was they had managed to get that first issue out. The bad news was that they had to do it all again in three months and find a way to pay for it. They did, and somehow kept it up for a year, clawing their way through by leveraging every cent available on their personal credit cards, and by getting a few relatives to put some charges on their cards as well. But they still needed more money, so Meghan went on a capital-raising mission that brought in $710,000 from angel investors.

The capital infusion kept the company solvent and it seemed that the dream was now a viable reality, but beneath the surface there were significant problems.

Most notably, there was a growing misalignment between what the magazine was about and how the company was operated. *Conscious Company* magazine gives business leaders information and advice based on three key principles:

1. *Conscious Leadership:* The self-awareness, wellness, and continuing personal development of the leader sets the tone for a conscious culture.
2. *Conscious Workplaces:* The workplace needs to be designed and managed so that it supports the wellness and continuing development of all employees.
3. *Impact:* The business must serve a purpose that goes well beyond generating a financial return.

Although they were doing a brilliant job of telling stories of leaders who were living these principles, the magazine's founders were drifting away from their own principles. Meghan was working seventy-five to eighty hours a week, not exercising, and neglecting her family and friendships. She and Maren gave no attention to the culture they were creating at work, and never had time to even think about taking care of their team. Their unspoken belief was, "We are doing great things in the world, the rest will take care of itself." They did not codify a purpose for the organization or the values and norms they would live by. Meghan recalls, "Despite the fact that we were doing great work and putting out a wonderful publication that people were finding hugely valuable—we had all these stories coming in from all around the world from people who started their companies as a result of picking up one of our magazines—internally, we were experiencing an unbelievable amount of discord and suffering."

The intense stress was manifesting in unpleasant disagreements about the direction for the company. They brought in a third partner in an attempt to create more harmony, but that only made things worse. By 2016, though it seemed to people on the outside that the magazine was thriving, Meghan found herself at the lowest point of her life. She was experiencing intermittent chest pains and full-fledged panic attacks and found herself in despair on an almost daily basis.

Moreover, despite the early success, the magazine was headed for bankruptcy. With money running out and the culture in disarray, the situation was dire. Meghan reflects, "Nobody was seeing eye to eye. We were writing about engagement but our team members were becoming disengaged. We were flailing around. I could not see an end to this. It was awful."

Meghan was distraught at the thought that she might disappoint so many people who had believed in her. "The idea that I was going to let down my investors (people who I had convinced to give me money), I was going to let down my team (people I had convinced to come work for me), I was going to let down our community (the people who read our magazine)—it was the most soul-crushing thing I have ever been through in my life."

It was clear that incremental changes were not going to remedy the situation; drastic action was needed. Meghan and Maren realized that if they could sell the company to a mission-aligned buyer, they could move forward and realign. After six months of working with a potential buyer, they sold the company on December 1, 2017, in a win-win for all parties, with Meghan remaining as CEO.

With this fresh start, Meghan thought about what she needed to do make the company true to the ideals it was spreading. How could she reset the company? How could she build a business that was healing? How could she build a compassionate organization, one that was in full alignment with who she wanted to be and how she wanted her team to feel? With the help of Nathan Havey, a consultant, Meghan and her team clarified the company's purpose, which is to:

Redefine success in business in service of all life.

They crafted a compelling statement of their core values, which are: treat people beautifully, choose joy and love, take pride in the product, practice radical trust and courageous patience, be authentic, stay open and curious, and walk the talk. They also came up with cultural norms and practices to adhere to on a daily, weekly, and monthly basis to ensure that they would stay true to their purpose and live their core values.

Meghan made a deep commitment to her own personal development. She got a coach and a therapist, adopted daily meditation and yoga practices, and started journaling.

The organization healed quickly from the trauma of the previous year and started to blossom. The new cultural practices helped greatly. For example, with a geographically dispersed team, the week begins every Monday with a virtual team meeting that opens with a personal check-in; people talk about how they're doing as humans and what they are most grateful for right now. The meeting ends by talking about the personal purpose of each of the team members (Meghan's is to be a force of love). They share their highs and lows, what they were proud of and not so proud of from the previous week in staying true to their personal purpose. On Fridays, the team focuses on the company's purpose and values and goes through a similar conversation about highs and lows. The company practices radical transparency, recognizing that anything people are going through at home will affect them at the office. Team members are able to bring any difficulties or traumas they are going through in life into the workplace and talk about them as a team. The focus is, "What do we need to do to support you?" The difference between the new culture and the old culture is profound.

The healing ideals that Meghan is championing at Conscious Company Media got tested ten days before we spoke with her. In the second month of her first pregnancy she suffered a miscarriage. In the face of this heartbreaking disappointment, she asked herself: "How do I lead in a moment of immense trauma? How do I show up on Monday and lead an all-staff meeting when internally my soul is crushed right now?"

She opened the next Monday morning meeting by sharing her experience with her entire team and asking for support. Her team rallied around her and provided a depth of love and support that made a profound difference in her recovery. She reflects,

> We've all experienced trauma, we've all experienced grief, we've all experienced loss, and we are going to as we continue forward in our lives. Imagine a world where the organization that you're either building or you're working within can be one in which you don't have to suffer alone, but you can actually go to your team for support. You can actually be in a place of caring and compassion when you go into the workplace.

In that first week after her miscarriage, Meghan took time off from work to take care of herself and recover her energy. By modeling self-care, she makes it easier for others in the organization to do the same when necessary.

Leaders can help create and shape healing cultures, and those cultures then help heal everybody, including the leader. Businesses can only rise to the level of consciousness of the leader. If leaders are in trauma, the organization will be in trauma.

Meghan reflects,

> As I get to know fellow founders and CEOs, when you actually peel away the veneer, there's so much suffering going on at the highest levels. Leaders aren't expected or encouraged to discuss their loneliness and anxiety. The pressures of being a leader are intense: You are responsible for the profitability of your company; you are responsible for your workers' well-being and safety; you are responsible for your own work. And, of course, on top of all that responsibility, being a CEO doesn't make you immune from life traumas.

Meghan is a natural hard worker who has cultivated the self-awareness needed to optimize her energy so she can bring out the best in herself and others. She observes,

I've learned that if I'm tense or anxious, if I am struggling through something, whatever it might be, that usually is a sign that I need to take a break . . . to do something restorative. That could be lying down for five minutes, or just petting my dog, or going for a walk or just writing in a journal for a little bit.

Meghan and her team are doing a much better job of living the ideals that they write about. They've also discovered that by contributing to the expansion and evolution of the Conscious Capitalism community, they are able to give and receive even more support.

Meghan reflects on why the work of Conscious Company Media is so important to her and, she believes, to the world.

Business is the largest aggregator of human potential on the face of the planet. It's the place where we spend the majority of our waking hours, and if we can actually have organizations that are addressing societal issues and also creating workplaces where people feel valued and appreciated, then people are going home and feeling a sense of excitement about their work rather than dreading it. That energy comes home and that spreads from homes out to communities and from communities out to the world. If everyone had the pleasure of being able to work at a purpose-driven enterprise, I think it would be one of the largest levers for change that we could possibly see in the world right now. Supporting the leaders who are behind those purpose-driven enterprises and making sure that they have the support and the care and the nurturing that they need is one of the most important things that we could be doing.

By integrating her healing ideals with her leadership of the company while now generating consistent profit, Meghan and Conscious Company Media are indeed redefining success in business in the service of all life.

HOW *DO* YOU GET TO CARNEGIE HALL?

HOW JABIAN CONSULTING TRANSFORMED THEIR BUSINESS MODEL TO INCLUDE FAMILIES AS STAKEHOLDERS

Sometimes in life, we don't realize the true cost of something until it is too late. A friend of ours was recently walking on a chilly December evening from Grand Central Station to Carnegie Hall to attend a concert. His "cocktail attire" wasn't offering him much protection from the frigid, biting wind, and cabs were impossible to find in the rush-hour swirl, so on a whim he flagged down a passing three-wheeled "pedicab" to take him the remaining ten blocks. Our friend was raised in India where he remembered similar three-wheeled contraptions, known as cycle rickshaws, as the cheapest form of transportation. He figured the short ride might cost him $10 or so, but was prepared to offer the operator up to $20 for braving the cold. When he inquired about the price, the operator mumbled something about charging by the minute. Our friend didn't think much about it, wrapped himself in a blanket, and sat back to enjoy the quick ride and the twinkling holiday lights. Upon arriving at Carnegie Hall, the operator pulled out his phone and started calculating. He said, "That was twelve and a half minutes at $6.99 a minute. You owe me $87.50." Our friend was outraged and exclaimed: "I am not going to pay that much for a twelve-minute ride. That is over $400 an hour!" He offered $20, but the driver protested and started becoming belligerent. After an exchange of threats to call the police, the operator snatched the $20, let out a torrent of choice New York expletives, and pedaled away.

Alas, this story is symbolic of what happens too often in life; we get taken for a ride and do not realize the cost until it's too late. This is the case with many of the most coveted professions in our society. The true cost, the human cost, is often unbearable. By the time we fully realize that, we are in too deep.

In the classic line about how one gets to Carnegie Hall, the answer isn't "a pedicab," it's "practice, practice, practice." The practice of management consulting has long been regarded as one of the most attractive career options for the top graduates of elite business schools. It is lucrative and intellectually stimulating work, offering relatively young people the opportunity to make a significant impact on the strategies and operations of major companies. But the glamour comes with a price: Most consultants must leave home every Sunday and don't return until Friday night. If they are lucky, they sometimes get to leave on Monday or come back on Thursday night. While this usually does not seem like a problem early in one's career, it starts to feel like a Faustian bargain after marriage, and can become a heart-rending burden once you have kids.

This was the story for Nigel Zelccr, Chris Reinking, and Brian Betkowski, the cofounders of Jabian Consulting, a unique management consulting firm headquartered in Atlanta, with offices in Dallas, Charlotte, and Chicago. It dawned on each of the founders that they had, without realizing it, put themselves on a path that wasn't in alignment with their life priorities.

They had each grown up with a notion of success that was based on prevailing American cultural norms in terms of status, position, and wealth. Chris's father, for example, was a successful technology executive but, like his contemporaries, he spent five days a week on the road. By Chris's senior year in high school, all that time away finally cost his father his marriage and strained his relationship with his family. Nevertheless, Chris grew up to become a consultant and found himself falling into a lifestyle with a schedule very much like his father's.

One change from Chris's father's era is that there are now many more women traveling for work. Chris married another consultant and they both thrived on their fast-paced lifestyles—until they started a family. That is when it hit Chris. "I looked around and recognized that continuing this way was untenable. I can't tell you how many times I sat in conference rooms at six or seven at night with fellow partners, who would call their kids to say bedtime prayers, or sing them lullabies. I thought, 'This is wrong. That's not the kind of father I want to be.'" Instead of coaching their kids in soccer or softball, he watched dads ask about the score of the game. He saw mothers visibly consumed with guilt as they spoke with their harried husbands at home trying to bathe and feed their small kids.

Chris thought to himself, "There has to be a way to do this work that we love . . . but also being able to value our spouses, our marriages and kids, and our community as a whole."

Nigel's journey was similar to Chris's. He worked for many years at Accenture and became a senior executive. Then he had his first child and his life and his priorities were transformed. Although he loved and excelled at the work, he began to question the personal price he had to pay to continue on this track.

Brian, who also had worked for Accenture, adds, "Growing up, and through my education at Georgia Tech, I always liked to build things. I wanted to know how things worked, and if something worked well, I wanted to find a way to make it better. That's the same mentality that helped us decide to start Jabian."

As high-end consultants, Chris, Nigel, and Brian were all skilled at helping clients come up with elegant solutions to their most important problems. One of the ways top consultants earn their substantial fees is by helping clients discover limiting assumptions that may be interfering with their success. They began to apply these skills to their own problem. The first thing that became clear was that the practices they found so painful are simply unquestioned industry norms; they were not handed down by divine decree. They realized that they could challenge industry dogma. The result was an epiphany expressed in the question:

Why not create a local consulting firm?

They decided to start an independent management consulting company with a life-altering difference: Consultants would never have to get on a plane for client engagements. *Never.* All their clients would be local in the cities where they lived. From their own experience, they knew that, by not flying and not being in hotels every night, they would have a lot more time to devote to things that really mattered, and they'd actually be in a better position to serve their clients.

Nigel states, "We changed the model and made this much more like a normal job; you get to go home at a normal time and sleep in your own bed." In addition,

No one objected to trading in the accrual of airline miles or hotel points in exchange for becoming a whole person. Our consultants now have much happier lives—they are soccer coaches for their kids' teams and they're active in community organizations, churches, and other local endeavors that enrich the quality of life.

Every year, *Consulting* magazine celebrates the top twenty-five consultants in the country, using a process whereby clients nominate the consultants who had the greatest impact on them.[1] Jabian is a small thirteen-year-old firm with a presence in only four cities, but three of Jabian's consultants have been honored in this way, competing against all of the big firms.

Kristine Jordan was one member of the Jabian team to receive this national recognition. A graduate of Texas A&M, where she majored in industrial engineering and minored in business. Kristine's father was a senior Andersen Consulting partner. Her mother ran a dance and gymnastics studio. Kristine became a successful consultant at

Accenture while also engaging her passion to practice and teach gymnastics (which she later gave up to teach dance). As she told us, "I literally became my parents—both of them!"

A rising star at Accenture, where she had worked nine years, Kristine was ready to start a family but began to feel that the life of a senior partner at the firm wasn't going to be a path to happiness or fulfillment. Her father had been a traveling consultant for twenty-eight years, and she knew what that life had meant for her mother and for herself. She could do what many of her friends had done, which was to leave consulting and get a regular job, or even stop working entirely. But Kristine loved management consulting and was superb at it. Fortunately, she met the partners at Jabian who were unanimous in wanting her to join the firm. After a period of careful contemplation, she decided to accept their offer.

Eight years later, Kristine has three children—including twins—and a happy marriage. She has been promoted three times. She is enthusiastic about the effects of the Jabian culture on her personal and professional fulfillment and growth. She says,

Soon after I joined, I realized that working here was making me be a better person. In the past, when I was traveling all the time, it was somewhat of a selfish view on life because I only had to worry about getting myself from point A to point B. I worked, I ate, I worked out, and I slept. Repeat. I was helping clients but I really wasn't contributing to the lives of friends or family.

Kristine adds,

The time I used to spend traveling is now time invested in offering much more in-depth help for clients and cultivating and enjoying relationships with colleagues and people in our community as well as having a wonderful family life. I've begun to appreciate the value of small kindnesses that brighten others' lives. Like the other day when I knew that a friend was having a rough time and I was able to stop and pick up a coffee for her on the way in to work. It's a small thing but it makes a huge difference, if you can bring that smile to somebody's face. It may be just one person, but there's a ripple effect. Maybe we are healing ourselves and others simply by being able to be present?

Unlike many larger competitors, Jabian employs a roughly equal number of women and men. By creating an environment that is conducive to family life, they can attract and retain stars like Kristine. As Nigel says, "Just think how many bright women have gone through college, are very talented, want to start a family, and yet the norms of the industry make it almost impossible for them to continue up the ladder. This is what they used to call the 'Mommy track.' Women had to step off their career path and get into a really slow lane if they also wanted to be mothers. These forced trade-offs were products of limited thinking and lack of caring. It doesn't have to be this way."

Kristine and her family are thriving in Jabian's family-oriented culture. They participate in a range of company-sponsored family programs including sporting events, picnics, bake-offs, and the annual Day of Service, held on Martin Luther King Jr. Day. She and her kids love to participate in the painting event sponsored by Atlanta's Foundation for Hospital Art. Kids three and older help "paint by number" several large canvases. Jabian works with several local organizations in each of its four cities on that day. Integrating the children into the day of service means that employees don't have to choose between staying at home with their kids, who are off from school, or participating. And, having missed much of his own children's childhood, Kristine's father is now making up for lost time by being constantly available for his grandchildren. He has earned the nickname Gruber, or Grandpa Uber, because he's always available to pick the kids up and do whatever else is necessary to support his daughter in her work and life.

At a company-sponsored family picnic, Nigel was approached by the husband of a new employee. He said, "I just need to let you know that I have never seen my wife have such a big smile. Since she came to work for you, she is happier than ever before."

In addition to jettisoning the road-warrior norm, Jabian has also disrupted another old-paradigm assumption of the big-scale firms: the practice of "up or out." In the big firms there are, every year, a limited number of promotions available. Those who fail to get promoted are asked to exit the firm. The result is often a cutthroat internal culture where people do not want to help each other succeed. In many firms the internal competition is much worse than the competition between firms.

Jabian has changed this in a way that is designed to promote a supportive, positive culture. They call it "grow or go." Rather than setting a limited number of promotion opportunities, Jabian invites people to demonstrate their commitment to grow personally and professionally. The firm employs an expert "career developer" to help all team members craft strategies to cultivate their potential and make the most of their opportunities. If you're not interested in continuous learning and ready to embrace a growth mindset, then this isn't the place for you! Unlike many firms, Jabian supports and rewards its people for their efforts in helping and supporting one another, and this pro-social, team orientation is a significant element in consideration for promotion at the firm. This positive culture is helping Jabian attract and retain many gifted consultants from larger competitors, who would rather work in a place where people really do care for one another, and want to succeed together.

In addition to the healing benefits for the consultants, their families, and communities, Jabian has discovered that being local and available is better for business. Chris explains, "Since we are not constantly flying off, we are in a position to develop deeper relationships with our clients." He offers the example of a vice president of technology at a major firm who was the internal lead on a project. Being local allowed Chris to get to know the VP and in the course of a few relaxed dinners to discover that he aspired to be a CIO. Chris helped him generate a plan and introduced him to his deep network within the local CIO community, and within six months the VP was promoted to CIO. As CIO, he was in a position to bring in management consultants and his choice was easy.

Jabian's pro bono work helped the state of Georgia develop its first-ever technology strategy, working with the governor and lieutenant governor to do so nearly ten years ago. Since then, Jabian has helped revamp and update the strategy. This is something that simply would not have been possible if they were operating as a conventional consulting firm.

Both of these are great examples of a win-win approach, because each also resulted in more business for the firm.

Nigel comments,

Many companies try to see what they can take out of the community. Our philosophy is how can you build up your community, because if you build up a really strong community, then we're just going to get more business from it. In weak communities, business starts to go away.

Nigel adds,

Before we knew about Conscious Capitalism or the Healing Organization, we used to just say, "We're going to do the right things because that will generate good corporate karma. We know that it feels good if we focus on the right things, and we have faith that the business side will work out." Even back in 2008–09, in the heart of the recession, we grew faster than ever because we just did the right things.

The old model of consulting imposes many human costs and delivers no additional benefits. The Jabian model offers all kinds of benefits without any added costs; in fact, there is a huge cost-savings because of all of the avoided air travel and hotel bills. Employees are healthier, happier, and less stressed. Spouses are no longer resentful and burned out. Children have a far better chance of growing up with more regular time and attention from the consultant parent. Local nonprofits get the benefit of pro bono work from some of the sharpest minds in the business world. Clients are able to enjoy deeper relationships with consultants who are available outside of the normal traveling consultant constraints. And all of this at no penalty to profits.

The business model for management consulting and other professional services firms was developed in a previous era, based on a win/lose, hypercompetitive, zero-sum set of assumptions. Law, advertising, and even medicine suffer from the same kinds of insane, dehumanizing notions about what it means to be a professional.

Jabian offers an inspiring example of what's possible when those assumptions are challenged, so that creativity, conscience, and compassion can be infused into business.

Frederick Winslow Taylor was one of the founders of the profession of management consulting. Taylor made his name when he met Andrew Carnegie at a gentlemen's club reception in Pittsburgh. Over Scotch and cigars, Taylor suggested that he had some management advice for the magnate, and Carnegie responded that he would pay $10,000—a vast sum in the nineteenth century—if the advice was useful.

Legend has it that Taylor's advice—that Carnegie should make a list of his top ten priorities and then focus on each, starting with number one—resulted in receipt of the first of many significant fees paid to consultants. Carnegie was one of the icons of American business who helped set the stage for how people think about success. He made a vast fortune, but in the process he damaged many lives and scourged the earth. He tried to make up for it by endowing libraries, universities, and, of course, Carnegie Hall. But now we know there's a better way to get there.

ENLIGHTENED HOSPITALITY

HOW USHG TURNS SIBLING RIVALRY INTO SIBLING REVELRY

Comedian George Carlin once quipped, "The other night I ate at a real nice family restaurant. Every table had an argument going." As is true with many healing leaders, Union Square Hospitality Group founder Danny Meyer's approach to business was shaped by his own challenging family dynamics.

In the classic *The Hero with a Thousand Faces*, Joseph Campbell taught us that we find our life's meaning and purpose by following our bliss, or our heartbreak, sometimes both. For Danny, the middle child of three, the bliss came much later, after he had experienced the turmoil and heartbreak of a difficult childhood in Saint Louis. Growing up during the Vietnam War, the Watts riots, and Watergate, there was an argument every night at the family dinner table between his liberal Democrat mother and his conservative Republican father. The disagreements were often heated and his siblings joined in, exacerbating the discord. Danny learned to listen and empathize with all sides of a discussion. It became important to him to find ways to bring people together and find common ground.

One thing that everyone agreed on at the Meyer family table was the importance of art, culture, and great food. When they weren't arguing, Danny's family was traveling to museums, historical sites, and wonderful restaurants in Italy and France, thanks to his father's entrepreneurial venture: an innovative travel agency that pioneered the idea of culinary tours to boutique European country inns. Unfortunately, Danny's father's business acumen wasn't on par with his passion, taste, and imagination and he eventually went bankrupt. And then his parents got divorced.

Danny reflects, "When I became a restaurateur at the age of twenty-seven, I actually mistook work for family. I think I was trying to create the family I wish I had in terms of respect, support, and belonging."

Of course Danny realized the difference: You can't fire family members, but sometimes you do have to fire people who work for you. But he also realized that, even when he needed to fire somebody, it could be done with care and respect. For better or worse, a restaurant really does feel like a surrogate family for many employees, because they work as a team, and most restaurant professionals spend more time with one another than they do with their own families.

When he started Union Square Cafe in 1985, Danny was following his bliss, his love of delicious, authentic, unpretentious food and artisanal wines served with grace and love. He learned from his father's business and relationship mistakes and also from the wonderful experiences his father provided, including the formative opportunity to dine at Taillevent in Paris, where he experienced the pinnacle of what a great restaurant can be. Danny resolved to create a positive, highly functional, healing family environment for all of his stakeholders.

He has succeeded beyond his wildest dreams. The Zagat Survey has ranked Danny's first restaurant, Union Square Cafe, New York's most popular restaurant an unprecedented nine times.[1] The cafe's success led to the evolution of Union Square Hospitality Group, spawning many other successful restaurants including Gramercy Tavern, Jazz Standard, The Modern, and the wildly popular Shake Shack chain.

The USHG restaurant family has won twenty-eight James Beard Awards. As two-time James Beard Award winner Karen Page, a Harvard MBA who travels throughout the country interviewing chefs and restaurateurs for her bestselling books including *The Flavor Bible* and *What to Drink with What You Eat*, attests, "New York's finest restaurateur, Danny Meyer, has had far-reaching influence, raising the standard of dining and hospitality both in America and beyond."

USHG now shares its wisdom on how to create great hospitality and offers consulting to other aspiring Healing Organizations. In his bestseller, *Setting The Table: The Transforming Power of Hospitality in Business,* Danny begins by sharing a prayer about the eleven restaurants he had opened at that point: "So far, I haven't had the experience of closing any of them, and I pray I never will." But after September 11, 2001, Tabla, his elegant, Indian-themed restaurant, began to struggle. When the 2008 recession hit, business declined further and he reluctantly closed the restaurant in 2010. More recently, at the end of 2018, a confluence of challenging circumstances led him to close his critically acclaimed Franco-American North End Grill in Battery Park. Despite these disappointments, his record of success in the intensely competitive New York restaurant business is astonishing: 75 percent of new restaurants fail, but nearly 90 percent of Meyer's succeed.

A huge element in this amazing success is the caring, healing environment he creates for his team, and that became even clearer when tested by adversity. Tabla's demise taught him a significant business lesson. As he explains, "You shouldn't take your most esoteric concept and fit it into the largest space with the highest fixed costs." He shares that on the day he told the restaurant's management team he had decided to close, "I don't think I've ever cried so hard in my life."

Meyer's loyalty to his team had inspired him to subsidize Tabla for two years with his personal funds, but it slowly became clear that the turnaround he was hoping for wasn't likely. His advisors convinced him that the artificial life support that he was providing wasn't going to revive the patient. "They argued that keeping a sinking ship afloat was the worst possible way to take care of the crew." But then he had an epiphany: Although the restaurant had to close, he could find openings for the people who worked there and position them to grow in their careers. "I started thinking, 'What if we could get these people more uplifting job opportunities either within or outside of our company?'"

This proved to be a great idea, as Tabla veterans went on to play key roles in a number of other successful Meyer ventures. As *Crain's New York Business* reported in 2010, Danny Meyer is "raising the bar on how to shutter a restaurant . . . in a field in which employees are lucky to receive a few days warning that they're losing their jobs, Mr. Meyer is going to extraordinary lengths to help Tabla's ninety full-time staff find new positions before the last meal is served on December 30."[2]

Meyer mused to the *Crain's* reporter: "It's not something I necessarily want to become an expert in. But the measure of our company should not just be about how we open restaurants. We also need to distinguish ourselves by how we close a place."

Danny's desire to create a healing family atmosphere at Union Square Cafe and beyond evolved into what he calls *enlightened hospitality*. We've discerned eight key elements of this approach.

1. START BY HIRING PEOPLE WITH A HIGH "HQ"

HQ, or *hospitality quotient*, is the first quality that Meyer looks for in recruiting and hiring his team members. He seeks people who naturally enjoy looking after others' needs. These are people who understand that "you feel better when you make other people feel good." Danny emphasizes, "Hospitality is a team sport," and people with a high HQ bring out the best in their team members so that together they can create wonderful guest experiences.

People with high HQ also tend to be more optimistic and to look for the best in others, and as we know from decades of research into the Rosenthal or Pygmalion Effect, this trait tends to inspire others to actually be their best, thus generating a virtuous cycle of positive expectations and high performance.[3]

2. BE AN AGENT, NOT A GATEKEEPER

For those of us who travel frequently and dine out often, arriving at an airline check-in counter, hotel desk, or restaurant podium, we recognize an immediate, profound, and vivid distinction between being welcomed as a human being or processed like an object. People with strong HQ are much more likely to serve as the customer's champion or agent. Those who don't have it tend to act more like what Danny calls "gatekeepers." He writes,

In every business, there are employees who are the first point of contact with the customers (attendants at airport gates, receptionists at doctors' offices, bank tellers, executive assistants). Those people can come across either as agents or as gatekeepers. An agent makes things happen for others. A gatekeeper sets up barriers to keep people out. . . . In the world of hospitality, there's rarely anything in between.

The agent/gatekeeper distinction is important as it touches something deep inside us. Meyer explains:

Within moments of being born, most babies find themselves receiving the first four gifts of life: eye contact, a smile, a hug, and some food. We receive many other gifts in a lifetime, but you can never surpass those first four. That first time may be the purest hospitality transaction we will ever have, and it's not much of a surprise that we will crave those gifts for the rest of our lives. I know I do.

3. INSPIRE EMPLOYEES WITH
A HIGHER PURPOSE, BEYOND TIPS

In the 1980 US presidential election, former Republican congressman John B. Anderson ran as an independent against Ronald Reagan and Jimmy Carter. Anderson inspired many college students and other volunteers with his moderate, environmentally prescient campaign. One of his supporters was the young Danny Meyer, whose experience organizing idealistic volunteers gave him a great insight into how to bring out the best in people. He states, "I learned to treat people as if they are volunteers."

If you start with people who love serving others and then give them the opportunity to develop and express their passion for hospitality, you generate an experience of fulfillment that redefines success. As Meyer notes, "If you're feeling like you're finding your balance between your spiritual needs, emotional needs, and physical needs, that, to me, is success."

Meyer backed up this principle with another radical innovation that he calls *hospitality included*. In Europe, where restaurant service has a long tradition of being a dignified and valued profession, the gratuity is "compris," whereas in the US, and especially in New York, being a waiter is something that people do while they wait for their big break on Broadway. Danny's decision to eliminate tipping created controversy and hasn't been easy to implement, but the goal is to benefit all employees through a more equitable and professional system of generating and distributing income, and to benefit customers by simplifying the transactional aspect of the dining experience. Judging by the growing popularity of his restaurants, it's working.

4. PROVIDE EXCEPTIONAL VALUE

Meyer also did something truly radical when he first opened Union Square Cafe: He didn't charge the same outrageously high prices that one usually has to pay for great food and service in New York. Since then, as diners have become more sophisticated, the demand for fine food and a welcoming atmosphere at a fair price has grown rapidly. Meyer has kept prices, including prices for the fine wines on his award-winning wine lists, at reasonable levels.

5. CONTINUOUS IMPROVEMENT AND SERVICE RECOVERY

Even when one of his restaurants wins a James Beard Award or is named best in New York, Meyer emphasizes, "There's always room for improvement." When he first started, professional restaurant critics wielded enormous power and could make or crush a reputation. Today, with Instagram, Yelp, etc., everyone is a critic. Meyer's policy is to listen and learn from all the feedback he can.

Acknowledging the challenge of maintaining consistent excellence and staying true to his high ideals, he told us, "I really try hard to believe that we can be this kind of healing organization, and yet I know that today there will be some human dynamic that does not feel healing to somebody. I think it is so important to have high ideals if you want to improve. I don't know that we fully live the principles every day, but having the ideals and listening to feedback gets you closer."

One element of healing that was part of the USHG culture from the beginning was a positive, responsive approach to what industry professionals call *service recovery*. As Danny explains, "A great restaurant doesn't distinguish itself by how few mistakes it makes but by how well they handle those mistakes."

In the early days of USC, there was a fair degree of chaos due to the demand for reservations, and because it was a learn-as-you-go endeavor for Danny and his team. Back in 1985, very few restaurants in the United States offered dessert wines by the glass. Danny kept a supply in a special refrigerator affectionately known as "the medicine chest." His team discovered that a complimentary Vin Santo from Tuscany proved to be the perfect remedy to sooth the *agita* of a customer who was kept waiting for a table.

Meyer also realized early on that the customer isn't always right. (Read his book to enjoy the compelling story of the time he was forced to punch out an unruly, abusive guest.) He reframes the question in an innovative way: "It's irrelevant who is right or wrong. What matters is that the customer must always feel heard. And the team member must also always feel heard. It is far more important for people to feel heard than to be right or to be agreed with."

He adds, "I have learned that it can be demoralizing to your team to say the customer is always right, because no one should have to compromise their own self-esteem or dignity to do something for the customer."

Standing behind employees in this way actually inspires exemplary customer service. Meyer explains, "Mutual support and respect for everyone's dignity and concerns contributes greatly to the healing environment. It promotes an ambiance that usually prevents the cause of most customer complaints."

6. LEVERAGE TECHNOLOGY FOR MORE HUMAN CONNECTION

Danny believes that technology must be leveraged to make the restaurant's job easier and the customer's experience more joyful. When Union Square Cafe reopened after moving to a nearby location, Danny equipped his managers and sommeliers with Apple Watches synchronized to a state-of-the-art online reservation system that al-lows the restaurant to store and access data on customer preferences. Danny exclaims: "Technology should be used to amplify your ability to use your heart!"

7. FACILITATE SIBLING REVELRY

Part of Danny's motivation in eliminating tipping from his restaurants was to promote cooperation and collaboration and reduce internal competition. At the Meyer family dinner table, there was lots of sibling competition for parental attention. Danny reflects, "Anyone with brothers or sisters has experienced sibling rivalry. The dynamic is very strong in humans and in the animal kingdom. Look at a bird's nest: When mom or dad comes back with one worm, and there's eight open mouths in the nest, there's going to be conflict."

Healthy competition, including internal competition, spurs people to improve, innovate, and grow. The question is: How can we ensure that competition remains healthy? Danny has thought deeply about this and has coined a phrase for his approach: *sibling revelry.*

The human dynamic of sibling rivalry goes back to Cain and Abel. Rather than seeking to eradicate it, which would go against the grain of human nature, Danny has sought to figure out how to harness that dynamic in a beneficial way.

> We are competing but we're competing to actually make the world a better place. We're competing to inspire one another by virtue of lifting each other up. Part of what I try to do as a leader and I try to teach other leaders to do is to actually celebrate internal competition, but in a way that you're genuinely happy for the success of your siblings, even though you'd like to actually have won.

An element of USHG's strategy to use this dynamic to the advantage of all stakeholders is its own self-administered Great Place to

Work survey. It has chosen not to use it to compete for recognition in the outside world, but rather to assess itself and hold itself accountable for improving the culture every year. Danny says, "We would rather compete with ourselves. And every single year we fall short of where we would like to be."

The real secret in creating a culture where managers, cooks, servers, and sommeliers genuinely support and celebrate one another's success is a strong emphasis on creating a culture of belonging. This manifests in an embrace of diversity and inclusiveness. Danny is passionate in his statement that "hospitality knows no gender or race."

During a recent opening of a Shake Shack in his hometown of Saint Louis, Danny was deeply moved by witnessing an expression of the fulfillment of his dream to create the sense of belonging he yearned for as a child. He explains,

Whenever we open in a new city, we bring a group of managers from other restaurants around the country to help with the opening. One of the managers had just been promoted last week and then got this opportunity in the first week to be part of the team that was helping to open St. Louis. As he was being introduced, the chant that happens in every Shake Shack when there is a new manager began—twenty people clapping and chanting: "One of us, One of us, One of us."

Danny shares,

I was so moved! I got goose bumps, because they were inculcating a primal notion of belonging, of being connected to something greater, for that manager. And reinforcing that solidarity for themselves, too.

This is a truly healing experience. In our society, many social institutions that used to fulfill people's need to belong have declined precipitously. Danny believes that the workplace can fill the gap in a way that is healing to society.

8. COMMUNITY HOSPITALITY

Danny's original vision for his first restaurant included a sense of community service. He foresaw that the Union Square neighborhood had great potential and he believed that he could attract customers who would then be drawn to help bring out the best in the surrounding area.

As soon as he became profitable, he began sponsoring charitable events for the community with a special focus on helping those suffering from hunger.

Danny's second restaurant, Gramercy Tavern, opened in 1994. A perennial fixture at the top of the Zagat Guide rating of Manhattan's most popular dining establishments, it's described as "about as perfect as a restaurant can get."[4] A recent Twitter post from the restaurant gives a sense of its healing culture:

> Today our team packed 2,800 meals for people too sick to cook for themselves in NYC. Thanks @godslovenyc![5]

Terry Coughlin began his career at Gramercy Tavern when he was just twenty-seven years old. He eventually moved on to help open Maialino, Danny's Roman-style trattoria, where he played an important role in establishing it as a profitable venture. Then Terry confronted the greatest challenge of his life: His younger daughter got leukemia. He explains, "I went through the hardest five months of my life, feeling so powerless with a child who was sick." Fortunately, she recovered. Terry was overwhelmed with relief and gratitude that translated into a passionate desire to help others. He went to Danny with a plan to leverage Maialino's popularity to raise money for a camp to support families with children struggling with cancer. Terry explains, "He didn't blink!"

Danny said, "Yes, let's do something special," and agreed to host an annual fund-raiser at the restaurant, and to buy a table personally.

Terry's initiative has thus far raised more than $2 million for the cause. He says, "I can't explain how moved I was that Danny gave his blessing and his real support to this dream. It means everything to me."

Terry adds,

This changed my life . . . being in the service industry takes on a whole new meaning. I was able to take the most horrific experience of my life and be able to heal on a level that I couldn't even imagine . . . when I realized I could help others, I was empowered and healed myself.

Danny Meyer's extraordinary success for over three decades is predicated on his healing orientation toward employees, customers, and communities. In New York City, perhaps the toughest restaurant market in the world—and a place not known for social niceties—taking genuine care of people and nurturing their spirits has proven to be the ultimate source of sustained competitive advantage.

Danny reflects, "In the end, what's most meaningful is creating positive, uplifting outcomes for human experiences and human relationships. Business, like life, is all about how you make people feel. It's that simple, and it's that hard."

FROM DEATH
MARCH TO JOY RIDE

HOW MENLO INNOVATIONS
CREATES GREAT SOFTWARE WITH
TWENTY-FOUR BABIES IN THE OFFICE

When Richard Sheridan was a little boy, his parents read him bedtime stories, including Aesop's Fables. His favorite was called "The North Wind and the Sun," which concludes:

Gentleness and kind persuasion win where force and bluster fail.

Richard Sheridan was blessed with an idyllic childhood, suffused with gentleness and love. Although his family didn't have much money, there was an abundance of joy and kindness. His parents had a conventional 1950s American marriage; she stayed at home while he worked. But the deep mutual respect and support they shared gave young Richard a clear sense of how people could play different roles with dignity and joy. "Our parents showered my two brothers and I with pure goodness and love and they imparted, by example and through encouragement, a passion for learning."

His parents encouraged his passion for learning about computers and information technology. He notes, "I wrote my first lines of code on a teletype machine in Mount Clemens, Michigan, when I was thirteen years old." He adds, "From what I know of the history . . . that was just a little over a hundred years after Edison was taught how to use the telegraph in the Mount Clemens rail yard!"

In 1971, Intel invented the first microprocessor, the 4004, and that same year Richard's high school began to offer a computer science class.[1] He became obsessed with coding at the perfect time, the dawn of the golden age of software. He got his first job as a programmer before he could drive a car. Richard was flabbergasted and thrilled that he could get paid doing something that he loved so much.

Eventually he graduated from the University of Michigan with degrees in computer science and computer engineering. His skills were in demand and companies competed to engage his services. He recalls, "Every year brought raises, promotions, and stock options with greater authority and a bigger office and all the worldly trappings of success!"

His zeal for coding continued, but he was dismayed to discover that the corporate atmosphere in which the work took place wasn't in alignment with the human values he had been raised to cherish. Although he was generally treated well since he brought so much value to his employers, he found himself surrounded by a culture that utilized force and bluster to manipulate behavior. The atmosphere was "crisis-centered, crazy, and dehumanizing." He explains, "The business was organized and managed so that every moment felt like a crisis. It wasn't unusual to work around the clock subsisting only on pizza and Mountain Dew. People talked routinely about a 'death march' toward a software project delivery date."

Although he valued working with intensity, the frenzied and unhealthy workplace led to results that compromised his instinctive commitment to excellence and quality. Many projects ran way over budget and a high percentage were, in a seemingly random fashion, canceled midstream. The software products that actually shipped were often riddled with defects, leading to unhappy users.

As he learned more about the industry, Richard came to the depressing realization that these dysfunctional practices weren't just true for the company where he was employed. Rather, "The whole industry seemed to have gone awry."

He began to wonder if there might be a better way. He started educating himself on the human aspects of business, immersing himself in the works of seminal thinkers in the realm of humanistic management and effective leadership like Peter Drucker, Tom Peters, and Peter Senge. He reflects, "I realized that, compared to human systems and relationship dynamics, technology was easy!"

Despite his unhappiness, he kept getting promoted. But his external success didn't assuage his internal dissatisfaction. By his mid-thirties, he could see that his path was unsustainable. Like most in the industry, he would routinely miss dinner with his family, coming home late, exhausted, and demoralized. In his twenties, this was unpleasant but tolerable, but now he realized that *his* life had become a death march. Feeling trapped and scared about the cumulative effects of the stress on his physical and emotional well-being, he was filled with dread at the thought of continuing in the same manner for another thirty years. His wife could see that his spirit was suffering and she encouraged him to apply all of his ideals and wisdom to changing his circumstances. After deciding that his fantasy of quitting his job to start a canoe adventure company in the boundary waters of Minnesota wasn't his true path, Richard asked himself the classic questions posed by Tom Peters:

If not excellence, what? If not excellence now, when?

Now was the time for him to embark on another kind of adventure: He resolved to apply everything he had learned about excellence, quality, and leadership from the books he had read—and, even more fundamentally, everything he had learned from his parents about kindness, respect, and the joy of living—to transform his workplace. He had become the vice president for R&D of Interface Systems, a public company in Ann Arbor, Michigan, and over the next two years he led an organizational transformation effort that remade the company culture. He changed everything about the way work was getting done, and rediscovered the joy that had drawn him to the software business in the first place. Life was becoming truly enjoyable again.

And then, in March 2000, the NASDAQ crashed and everything started to melt down. Although the bursting of the dot-com bubble seemed like a disaster for him, it turned out to be a blessing. He couldn't continue his enlightened leadership practices at Interface: It was acquired by a Redwood City, California, company in September 2000 and then shuttered in April 2001. This created a void that he could only fill by starting the company of his dreams.

First, he needed a name, and that was easy. Besides his parents, Richard's inspiration as a child came from his hero, America's greatest inventor, Thomas Edison. When he was eight years old, his parents took him to Greenfield Village, a Detroit-area living history museum that features the reconstruction of Edison's Menlo Park laboratory. Edison's philosophy, as expressed in these three quotes, influenced Richard's approach:

- If we did all the things we are capable of doing, we would literally astound ourselves.
- There ain't no rules around here! We are tryin' to accomplish something!
- I never did a day's work in my life, it was all fun.

Edison is remembered as the seminal figure in illuminating the planet, originating the phonograph—and therefore the contemporary industry based on sound recording—and inventing the technology that launched the motion picture industry. But Richard understood that, although Edison holds the record of 1,093 individual United States patents, his *greatest* invention was the world's first collaborative process for systematic innovation.[2] Richard and his partners wanted to create a collaborative culture of systematic innovation without conventional rules. They wanted to astound themselves and their stakeholders with their achievements and do all this in a way that was fun and filled with joy. So they named this new venture Menlo Innovations.

From the beginning, the Menlonians, as they like to call themselves, had a clear and powerful healing purpose:

End human suffering in the world as it relates to technology.

As Richard expresses it:

Creating software is one of the most unique and useful endeavors undertaken by mankind. There's a fundamental joy and delight in this creative process and we were determined to build a workplace that highlights this.

Richard saw an opportunity to alleviate the unnecessary suffering that was just taken for granted by each of the company's three major stakeholder groups.

First are the people who pay for the software to be built, who get upset and angry when budgets are overrun and schedules are missed and quality is absent.

Second are the end users, the people developers never meet, but whose lives are impacted for better or worse by the code they write. Richard says, "Software typically tortures people. As an industry, the people served are known as 'stupid users,' so 'dummies' books are written for those poor people. Defects and glitches are renamed as 'features.' It doesn't have to be that way. We decided from the beginning to honor the people we serve. We invented an entirely new practice to do that, what we call high-tech anthropology."

The third, and most pivotal, group are the people who create the software. Richard realized that the key to solving the first two issues was to take great care of his team. Menlo Innovations sought to address two of the main causes of suffering for employees in the industry: overwork and isolation. They do this first by adhering strictly to forty-hour workweeks. No more death marches! And they created an environment in which every developer works with a partner, rotating to a new partner every five days so that everyone learns to collaborate with everyone else and no one is ever stuck with a project on their own.

This rotating-partner, collaborative approach changed one of the dysfunctional habits in the tech industry: a dependence on the "tower of knowledge," the one person on the team who knows everything about something, way beyond what anybody else knows. That can feel good to the person who is so regarded, but eventually it becomes a prison they cannot escape. They are held hostage at the office and their requests for vacation time are often denied, since they are irreplaceable. Instead, Menlo created an interdependent system, in which knowledge sharing is valued and encouraged. Pairing and switching the pairs turned out to be a brilliant way of accomplishing this knowledge transfer every day.

Richard reports that Menlo hasn't had a serious software emergency in fifteen years, in an industry that usually lurches from one emergency to the next. "We don't have a need for firefighting because our collaborative, balanced approach prevents fires from breaking out in the first place." He adds, "By limiting work to forty hours per week and encouraging our people to take full vacation time, we've nurtured a group that works much more efficiently and effectively."

Menlo Innovations has become a place that aspiring entrepreneurs and established businesses all come to visit to learn about the revolutionary path to a joyful and profitable culture. A delightful evolution of that culture began in 2007 when Tracy, who had been working at Menlo for just a year as a quality advocate, became pregnant with her second child. Menlo already had a generous maternity leave program, but after she took time off to have her baby and felt ready to come back to work, she found that the day care she had planned to use was full. Her parents and in-laws lived too far away to help. When she called Richard to explain her predicament, he responded, "Bring your baby to work."

That was, as of this writing, twenty-four Menlo babies ago. Babies come in with mom or dad, sometimes both, when a couple works together. Richard says,

> We have a wonderful setup here to support the parents and it is a pure delight. It blows people away that we do this. The joy and delight the babies bring to the office is priceless. The team helps raise the children! Even our customers behave better when there are babies in the office.

As the babies keep coming, Menlo gets better and better at integrating infants into the workplace. For example, Menlo received a gold medal award from Washtenaw County Breastfeeding Coalition for their innovative policies to support breastfeeding on the job. As technology experts, they had no trouble installing monitors so parents could listen for sounds of disturbance when the babies are sleeping in a separate room. When they're awake, the babies are playing around the office, suffusing the environment with joy and laughter, especially when they make sounds that Richard describes as "dolphin meets pterodactyl."

Menlonians are also free to bring their dogs to the office, and this pet-friendly policy multiplies the positive energy all around. On snow days, when Ann Arbor schools are closed, employees bring their children to the office and conference rooms get converted to movie and play rooms. The ambiance is magnetic and enlivening and attracts customers who sometimes bring their own children when they visit. A customer once brought his daughter with him to the Menlo office and told Richard, "I knew I didn't have to ask if I could bring her. I knew it would be okay. That's just the way you guys are."

Menlo was born in the aftermath of the dot-com collapse, and soon was hit by 9/11. Two wars and the Great Recession followed. Through it all, the company has grown steadily and is successful consistently. Why? Richard explains: "We have extremely happy employees who create great software." This results in long-term relationships with clients.

Based in the high-cost city of Ann Arbor, Michigan, Menlo utilizes relatively expensive US, Michigan-based talent. It doesn't offshore anything. It doesn't outsource anything. Richard and his team offer a compelling case that helps potential clients calculate the cost of software-based emergencies, and the opportunity cost of failed projects. Once these calculations are clarified, Menlo customers are happy to pay more per hour, versus discounted rates available offshore, because they recognize that reliability saves money over time.

Many people take functional software for granted, but if it were easy, we would not have major disasters such as Equifax, or healthcare.gov, or airlines canceling flights all over the world because their systems are down. Software quality and reliability really matters, and savvy clients know they can count on Menlo. Richard says, "Sad, stressed, tired people make bad software, and bad software makes people more stressed, tired, and unhappy. We want to make great software, so we have happy, joyful, and high-energy people produce it."

After almost two decades in business, visitors to Menlo often *still* say to employees, "Wow, you're so lucky, you get to work for a startup!" Menlonians' exuberance, their childlike joy and innocence, is contagious. It inspires the stream of visitors who come to learn the company's secrets.

They discover that there are no secrets. Menlo shares everything they have learned about creating a culture of healing and joy. Beyond the babies, puppies, and forty-hour workweek with generous vacation time, Menlo has also evolved other cultural practices that promote collaboration and high performance. They dispensed with the usual corporate hierarchy; there are no bosses and no reporting relationships. All hiring, firing, and promotion decisions are made by the team. The company practices open-book finance. Everybody knows everybody else's compensation, and the teams are in control of it.

Richard is quick to remind visitors that Menlo arrived at this approach through an organic process of communication, soul-searching, and evolution. His goal is not to offer a formula for others to copy but rather to inspire others to discover how they can chart their own authentic journey toward healing, joy, and success.

In the bio at the back of his most recent book, *Chief Joy Officer: How Great Leaders Elevate Human Energy and Eliminate Fear,* Richard describes his position as "Chief Storyteller." He explains, "We do a lot of tours in which we tell many stories of how we got to where we are. We've been through pain and even desperation but managed to stay true to the guidance of our hearts." Richard realized that sharing stories of goodness, generosity, creativity, and courage in the face of adversity was the most important role he had as a leader. He en-

thuses, "Throughout history, civilization and culture have been cultivated through storytelling." The Menlonians are rewriting the
narrative for the software industry, from the cold wind of the lonely,
stressful, and isolated death march to the bright, warm sun of a collaborative, systems-oriented joyride.

FROM ELEGY
TO EXULTATION

HOW INTERAPT IS BRINGING DYNAMISM
AND DECENCY TO HEAL COAL COUNTRY MALAISE

J. D. Vance, author of *Hillbilly Elegy: A Memoir of a Family and Culture in Crisis*, was born in Kentucky. He grew up in a town that "has been hemorrhaging jobs and hope for as long as I can remember."[1] His bestselling memoir explores the complex cultural, racial, and economic dynamics underlying "rust belt" malaise, as manifested in some of the country's highest rates of alcoholism, opioid addiction, obesity, and domestic violence. Vance's book generated controversy and was interpreted quite differently by conservatives and liberals, but practical solutions to the malaise were not forthcoming from pundits on the right or the left.

Ankur Gopal was also born in Kentucky, the son of immigrants from India. His father came to study engineering and then got a job as a petroleum engineer in the heart of coal country. He returned to his homeland for an arranged marriage to Namita, a physician, and brought his new wife back to Kentucky. The young couple worked hard to assimilate, and were eventually embraced by their local community. Namita started a free clinic serving thousands in need and became a beloved figure in their small town. When Ankur was sixteen years old, he was checking out at Walmart when the cashier asked him, "Are you Dr. Gopal's son?" He replied, "Yes, that's my mom." The cashier said, "Your mom took care of us when no one else would. She got me off drugs and back into school, then she helped me get a steady job. Our family is thriving now because of her. We love your mom so much that we named our daughter Namita!"

Ankur's father taught him diligence, hard work, and duty and his mother taught him caring, loving-kindness, and compassion. He recalls, "I grew up like a typical Indian kid in the Midwest. I was good at tennis, school, and very good at spelling!" Like most bright kids from Kentucky, he thought success meant leaving the state. So, after graduating from the University of Chicago, he worked for Accenture in Silicon Valley for five years, then became an entrepreneur, starting and then selling a technology-oriented health-care business. After that success, he began to focus on another startup idea to leverage his software expertise in the health-care space. Fully intending to return to Silicon Valley, he incubated the new company, called Interapt, while staying with his parents in Kentucky.

When local business leaders urged him to stay in the region, his initial response was, "That's ridiculous." Kentucky's economy was struggling with the decline of the coal industry, and it seemed an unexciting and depressing place to be. But Ankur eventually became convinced that the region needed as many entrepreneurs as possible. He could see the value of being one of just a few such companies in Kentucky instead of being one of thousands in Silicon Valley. Interapt grew steadily and Ankur was honored with many awards, including the prestigious 2016 EY Entrepreneur of the Year Technology Award.

Although his achievements were already extraordinary, his conscience compelled him to want to do more to help alleviate the suffering in his home state. That's when something amazing happened. He received a call from the state's governor, Matt Bevin, and the congressman from his district, Hal Rogers. They said, "Eastern Kentucky is hurting. . . . Can you help us create jobs and opportunities?"

Inspired by this request, Ankur conducted careful research. He made three key recommendations:

1. "Economic development demands workforce development; people need to learn new skills."
2. "Barriers to learning new skills must be removed, which means we need to pay people to learn."
3. "We have to instill hope; that's one thing that's missing here."

Ankur created a training plan for people from the local community. He already had a program at Interapt to train his mostly college-educated employees to become software developers and coders. He had never done it with people transitioning from a displaced industry or with only high school educations. But he was inspired by the challenge.

He received eight hundred applications for the fifty places available from locals who had no software background, but who were hungry to make a better life for themselves and their families. The four-month training was followed by a two-month apprenticeship. It was rigorous, with classes eight hours a day, five days a week; if students passed the exams, they got to work on projects in a real software development life cycle alongside senior Interapt engineers.

Instead of charging for the training, Interapt paid participants $400 a week. After six months, thirty-five trainees out of fifty in the initial cohort were placed into technology jobs. Before the training, their salaries had ranged from zero to $17,000 a year. After the six-month program, they were earning $30,000–60,000 a year. Interapt extended job offers to twenty-five of the participants, but many of them had multiple job offers.

Like his mother, Ankur is trying to heal wounds in the culture. Depressed communities aren't going to be revitalized by false promises to bring back coal mining or by social welfare programs that compromise human dignity and stifle initiative.

Efforts like this are needed a thousand times over. Ankur says, "Historically, education has not been much valued here. For a hundred years, the joke was that if you went to high school, you made $100,000 a year, and if you went to college, you made $40,000 a year. But that's just not true anymore. We have to help people change their mindset. Looking backward at sunset industries isn't going to help. But there's a new sunrise and we are committed to helping our neighbors participate in it."

Interapt is now extending its offerings to transitioning soldiers and their spouses to help them assimilate into civilian life with new skills and a promising career. Ankur is also working with the "City of Refuge" in Atlanta, where the Interapt Tech Academy is giving twenty-five young people—many of whom live in shelters and are stuck in a cycle of poverty—new hope and new skills for a brighter future.

Ankur's goal is to train ten thousand people in the next five years, and he is rallying to respond to inquiries from around the country to replicate his program in other states as well as other regions in Kentucky.

Interapt is a profitable company, but it is much more than that: it has transformed lives. An unemployed single mother of two completed the program and has now become a member of the development team. On weekends, she teaches her son and daughter coding so they can get a head start. An ex–coal miner had been told he was too old to learn a new skill at forty-three and was hard-pressed to get a job bagging groceries. Now, recruiters contact him several times a month on LinkedIn to ask if he is open to other opportunities. When a trainee finished the program and got his first job, he and his wife posted on Facebook that it was the proudest day in their family's history since they didn't need to be on food stamps anymore.

Ankur reflects, "When you give people an opportunity, what you're doing is you're telling your fellow human beings that you belong, you are worthy, you are relevant, you have hope, and you are enough." Country music legend Willie Nelson, who has been called "the Hillbilly Dalai Lama," proclaims, "When I started counting my blessings, my whole life turned around." The son of Indian immigrants and his colleagues are blessing their communities with jobs and hope and turning around the narrative from elegy to exultation. Ankur exults, "Kentucky coal powered the nation for the past hundred years. Kentucky code will do the same for the next hundred."

THE SPIRIT OF GIVING

**HOW H-E-B PARTNERS SERVE
THEIR COMMUNITIES AND DEMONSTRATE
HEROISM IN THE FACE OF DISASTER**

H-E-B is a privately held grocery retailer based in San Antonio, Texas, with more than 110,000 employees, known as Partners. Operating more than 400 stores in Texas and northern Mexico, the company generates more than $26 billion in annual revenue with over 5 percent of pretax profits going to charity.[1]

The name H-E-B is derived from the initials of Howard Edward Butt, who assumed control of the small grocery store founded by his mother when he returned from serving in the Navy during World War I, in 1919. In 1905, with her husband disabled by tuberculosis, and with three small children, Florence Butt had begun selling grocery items that she acquired with her life savings of $60. She did this in a tiny 760-square-foot space—family living upstairs and store downstairs—that she was able to rent for $9 per month.

A courageous, pioneering entrepreneur, Florence realized that delivery service would help sustain her new business so, as the story goes, she loaded up Howard's baby carriage with food and went door-to-door.

As the store became viable, Florence and Howard delivered food and other essentials in his little wagon to homeless people camped on the banks of the Guadalupe River in Kerrville, Texas. From the beginning, the family's character centered around a belief they called the Spirit of Giving.

Despite some early missteps, Howard persevered and successfully opened forty stores in his fifty-two years at the helm. In 1971, his son Charles became president and remains the chairman and CEO. A beloved leader, Charles appears consistently in Glassdoor's ranking of the top ten CEOs in the United States.

Like all Healing Organizations, H-E-B takes great care of their Partners, vendors, and customers. As they like to express it:

We're in the people business. We just happen to sell groceries.

H-E-B partners love their company: If you ask a partner what H-E-B stands for, they'll say: Here Everything's Better.

Perhaps the most distinguishing and extraordinary aspect of the healing H-E-B brings to Texas and the world is its disaster relief operation. Few companies ease hardship on H-E-B's scale, and no company, that we know of, is more effective in mobilizing its resources to help when disaster strikes.

On August 25, 2017, Hurricane Harvey, a category four storm, punched the state of Texas hard, right in the gut. It dumped 60.5 inches of rain in some locations, a record for a single storm in the continental US. The weight of a foot and a half of floodwater, covering nearly 1,300 square miles, in which 4.5 million people resided, actually depressed the earth's crust; the city of Houston sank by nearly an inch. Katrina-like in its destructive ferocity, the storm damaged 204,000 homes and destroyed more than one million vehicles.[2] Three quarters of a million Texans registered for assistance with the Federal Emergency Management Agency (FEMA). As the storm approached landfall, government and philanthropic aid agencies worked frantically to mobilize relief efforts. Many companies were making their own preparations, but none with more know-how, generosity, love, and determination than H-E-B.

Committed to always being the "last to close and first to open" in the face of disasters, H-E-B maintained a fully staffed command center at the company's headquarters in San Antonio, which was complemented by command centers in Corpus Christi and Houston, a focal point of the storm. In addition to keeping as many stores open and stocked as possible, the command centers directed and coordinated relief efforts to the community.

As the storm approached, H-E-B started shipping huge amounts of milk, bread, water, canned meat, batteries, and other essential supplies into the affected areas. The company deployed its three fully equipped mobile kitchens, which can each feed over two thousand people an hour, directly into the heart of the hurricane area to serve first responders and evacuees. It also sent its mobile Disaster Relief Units that include a fully equipped pharmacy as well as a business center allowing customers to cash checks, receive money orders, and use ATMs.

A manifestation of its Spirit of Giving philosophy, H-E-B has a long history of dealing with natural disasters within its service region and beyond. H-E-B donated ten truckloads of vital supplies to Florida-based supermarket chain Publix toward relief efforts for Hurricane Irma. They sent several planeloads of goods to Puerto Rico in the aftermath of Hurricane Maria and contributed more than 350 tons of food to survivors of the devastating 2017 Puebla earthquake in Mexico.

But Hurricane Harvey presented a challenge of unprecedented magnitude. Many employees were displaced by the flooding and could not get to work. A typical store employs three hundred people, but in the storm conditions some stores operated with as few as five people, one at the door and four at checkout stands. The company used helicopters to fly truck drivers in from San Antonio, so they could get trucks out of the yard and to the stores so there would be groceries on the shelves. More than two thousand Partners from Austin, San Antonio, the Rio Grande Valley, and other locations answered the call to volunteer at stores in the areas under the most duress, and to help with a wide range of community relief efforts. Many just hopped in their cars and drove to the Houston and Gulf Coast areas where they worked long days and slept on couches in the homes of other local volunteers.

Justen Noakes, H-E-B's director of emergency preparedness, received a phone call a few days after the hurricane hit from the emergency management coordinator (EMC) for the city of Beaumont, near Houston. Inundated by floodwaters, the city had lost its tube pumps and had no water pressure and therefore no usable water. FEMA, the Red Cross, and the state of Texas were trying to help but their resources were stretched, and they couldn't respond. The EMC called Justen, saying: "I know that I'm asking a lot, but what can H-E-B do to get water to the citizens of Beaumont?" A convoy of water tankers was mobilized immediately with H-E-B's senior vice president of supply chain in the lead truck. A trip that normally takes ninety minutes took eighteen hours through two feet of water. Still, by the next morning, residents of Beaumont had clean drinking water, which prevented a public health crisis and the widespread panic that would have followed.

Scott McClelland, one of H-E-B's senior executives, recalls the heartfelt expressions of gratitude and appreciation from customers who were keenly aware of the effort involved for H-E-B to remain open for business under the circumstances, especially as other retailers were shuttered. McClelland says, "As soon as I entered the store, a woman walked up to me and started crying as she hugged me to thank us for being open."[3]

H-E-B's amazing *esprit de corps* is expressed when disaster strikes, but it's cultivated daily in the everyday management of the operation. When Craig Boyan was named the president of the company, Charles Butt told him and his leadership team: "Pay our people as much as you can, not as little as you can." Charles explained, "The income gap between top and bottom earners is too great for the nation's future stability."[4] Boyan comments, "We try to be the most competitive retailer we can possibly be. But we also try to be a great company, so we can take care of our Partners and their families. We pay better wages and benefits than most other retailers." He adds, "We are constantly looking to push wages up."

On its 110th anniversary in 2015, H-E-B announced an employee ownership plan that will eventually lead to employees owning about 15 percent of the company.[5] When the company launched the plan, 55,000 Partners received stock. Most H-E-B Partners are enrolled in the plan. The goal is to help partners with their retirements by supplementing their already generous 401(k) plans.

Beyond the financial rewards, H-E-B's culture brings out the best in people by creating a shared sense of purpose and meaning. They do this in a down-home Texas style. Indeed, a major element in their popularity is leveraging the pride people have in their state. H-E-B brings forth and celebrates the diversity of Texas culture and offers all kinds of promotions with the San Antonio Spurs and other state athletic teams, and by featuring Texas regional food specialties— don't be surprised if you're handed a free tortilla as you shop, and they even sell a Texas-shaped frying pan!

Craig Boyan explains, "We take ordinary people doing ordinary jobs and help them do extraordinary things and make them proud of being extraordinary retailers." For example, the company has its own trucking fleet and celebrates "greatness in trucking" by inducting drivers who have driven one million consecutive safe miles into its Truck Driver Hall of Fame. The company is also a leader in sustainability practices and was named to the *Forbes* list of America's Best Employers for Diversity in 2018.[6]

Caring for Partners translates into benefits for customers, and customer loyalty translates into consistent profitability that is then shared with Partners and the community. Boyan says, "Our customers can really feel that we care about what's important to them. We aim to meet the needs of all the different market segments in Texas." For this, H-E-B has a successful foodie format called Central Market, several Latino-focused markets named Mi Tienda, Joe V's Smart Shop, a value format, and many other initiatives designed to thrill their customers.

Boyan adds, "We aim to offer the best products, including some of the most unique products, at the lowest prices in the country."

In addition to their extraordinary disaster-relief efforts, H-E-B is one of the largest philanthropic givers in the state of Texas, supporting a range of worthy causes including:

- H-E-B Excellence in Education: Charles personally donates heavily to public education, and H-E-B contributes more than $10 million annually to education-related programs in Texas, including awards honoring outstanding public school teachers, principals, and superintendents. It also supports diversity and inclusion initiatives and early childhood literacy programs.
- H-E-B Food Bank Assistance: Since its founding in 1982, this program has delivered more than one billion pounds of product to seventeen food banks in Texas and Mexico.
- H-E-B Feast of Sharing: These are holiday dinners that are part of the company's efforts to combat hunger. The company annually serves 33 holiday meals to more than 250,000 people throughout Texas and Mexico.
- H-E-B Tournament of Champions: One of the largest charitable events of its kind in the nation, it has raised over $100 million for over six hundred Texas nonprofits focused on youth, education, and military families.
- H-E-B Operation Appreciation: This was launched in 2013 to honor veterans and active-duty service members of the US Armed Forces. Among its many efforts, the program has gifted more than twenty-five mortgage-free, custom homes to wounded warriors and their families.

For H-E-B, caring for the community isn't an afterthought, like many corporate social responsibility efforts. Rather, it is a manifestation of the giving spirt of founder Florence Butt, going strong nearly 115 years after she brought this extraordinary company into existence.

THE MEANS
AND THE ENDS
ARE THE SAME

HOW SOUNDS TRUE UPLIFTS THE
WORLD BY DISSEMINATING SPIRITUAL WISDOM

Florence Thornton Butt, the founder of H-E-B, was a devout Christian from Tennessee who moved to Kerrville, Texas, when her husband got tuberculosis and couldn't work.[1] Both her brothers were pastors and she volunteered at her church on a regular basis. Florence opened her little grocery store and prayed that the Lord would bless her enterprise. Her prayers were answered beyond what she might have imagined.

Eighty years later, another pioneering businesswoman, Tami Simon, founder of Sounds True, offered a prayer from the depths of her heart that led to the creation of an innovative multimedia business that has touched the lives of millions. Tami was raised in a much different cultural and religious milieu than Florence Butt. Her liberal Jewish family went to temple every Friday night, but more from a social-culture motivation than a religious one. Neither of her parents believed in God; they were post-Holocaust Jews who wondered: "How could God have let this happen to us?" But Tami was a spiritual seeker from a young age. As she recalls, "My mom was rational and kind of suspicious of anything metaphysical and my father was sleeping in his chair, but I was praying my buns off, really into it."

Tami went to Swarthmore College as a religious studies major, to explore the teachings of the great spiritual traditions of the world. Dissatisfied with what she perceived as a sterile, academic approach, she yearned for direct experience of transcendent knowing. So, she dropped out of college and took off for Sri Lanka, India, and Nepal for a year, learning meditation and other practical methods to awaken consciousness. Although far from her physical home, she felt as though she had discovered her spiritual abode. Tami explains,

> When I started meditating, I felt like I had found a gateway, a method, and it was what I needed. I made a commitment to introduce as many people as possible to the practice of meditation, and, more widely, to any method that helped people tune in to their own inner knowing through direct contact with a sense of aliveness, purpose, and inspiration.

Although her metaphysical quest was successful, her incarnation was compromised. Like many Western seekers who went to India and Nepal, Tami contracted hepatitis. She was dangerously underweight and realized that it was time to come home to the United States. She reflects,

> I had experienced something that was infinite and timeless but I had no idea what to do with it. I had hepatitis and weighed ninety pounds. My parents were worried. They were especially alarmed because I wasn't talking, and I was known as "Big Mouth" when I was a kid.

As she recovered her health and her loquacity, Tami started to work as a waitress and also volunteered at a local public radio station in Boulder, Colorado. Although her journey to the East gave her a taste of authentic wisdom, and she knew that she wanted to share what she learned, she also felt guilt about dropping out of college. There were no job openings or clear career paths for someone with her background. At the same time, she felt gratitude for all the opportunities provided by her family, who loved and supported her despite her deviation from the path they might have preferred her to take. She was grateful for everything she learned on her journey and felt a deep desire to contribute to helping others and healing the world.

From the depths of her being a prayer emerged:

God, I'm willing to do your work. Please show me what it is.

Tami repeated this prayer over and over, day and night.

Then her father died, leaving her a modest inheritance of $50,000. Although still waitressing to earn money, her volunteer work at the public radio station had led to the opportunity to host a show in which she interviewed a variety of spiritual teachers as a way of continuing her own learning and sharing the exploration with others.

One of her guests was a wise teacher who was also a successful entrepreneur. Tami had an intuition that he might have some ideas that would help her, so she arranged to meet with him at his office. Tami told him about the bequest she received and her question about what to do with it. He told her, "Wherever you put your money, that's where you're putting your energy." She replied, "Well, I don't know where I want to put my energy." He said, "Yes, you do. Come back in three days and we will talk about your next steps."

Tami's next steps, literally, changed her life:

When I walked out of his office, I had a strange experience: I felt like I was walking slightly off the ground. I thought, "Oh, my God. I'm not exactly in my body. I'm walking on air."

Then she heard an inner voice that said: "Disseminate spiritual wisdom." Tami adds,

It was more real than real. Nobody could take it away. It was a clear felt sense: "Yes, I will do this. It's been given to me as my operating instructions."

As Tami continued to walk along the streets of Boulder, her feet seemed to return to the ground and she began thinking in practical terms about how she could follow her new operating instructions. She realized that her volunteer radio program could be the beginning of a business. At college, she had discovered that she liked listening to lectures more than she enjoyed reading. Besides, she reasoned, book publishing was an established industry and there were already imprints specializing in spiritual wisdom. Video didn't interest her. As she explains, "I hated television at the time. My parents watched a lot of TV and I thought that it dumbed down conversation." The path forward became clear. Tami thought, "I'll start by disseminating spiritual wisdom with audio. That's a good medium for me. I love listening. I love the radio and I already have a radio show."

Tami was pleased to discover that listeners wanted copies of her interviews so she began making cassette tape recordings and selling them for ten dollars each. At first she sold about three copies in a week, ten in a really good week.

Despite her complete lack of business training or experience and with what she describes as "only a vague idea of what I wanted the company to be," Tami now had a clear mission and purpose that was in harmony with her deepest instincts and her true nature. She loved audio and she loved the quest for truth. The perfect name arose effortlessly: *Sounds True.*

"My biggest instinctual gift," states Tami, "was always just my love of the truth, a love of genuine spiritual teachings. That was just in me as a person." Tami's instinct for truth was evident from the beginning. When her parents watched politicians on TV and Tami was in the room, she would ask precocious questions like: "Why is he lying? How can that person be a leader? And why do people watch someone who is not telling the truth?" Tami explains, "I could always see when there was something that was out of sync between what was happening inside the human and what they were saying. It outraged me."

Tami's distaste for incongruence, dissimulation, and subterfuge motivated her to work with and record people with "a congruence between their inner life, their work, and their vocal expression." She adds, "I love when I'm with somebody who has that kind of congruence. I relax, I feel safe, and feel like I can come forward. What you see is what you get in them, and what you see is what you get in me."

Tami's candor and willingness to speak straightforwardly didn't always lead to acceptance and popularity. She admits, "I don't think my genuineness has been particularly well received, historically. I have felt like an outsider in many situations. For a long time I felt kind of like an alien. I wondered, 'Maybe I came from some different star and somehow got into a human body, I don't know how.'"

She adds, "Part of the reason I think I ended up starting my own business was because I didn't think there was some existing business that I could plug into." As she's matured, Tami has refined what she calls her "edit function" and learned to be more diplomatic without compromising authenticity. She's integrated her values of kindness and compassion with her fealty to truth and this has become the basis of her effectiveness as a leader.

Tami embodies what poet and Jungian analyst Clarissa Pinkola Estés refers to as the wolflike or wild nature. Estés writes that when women are in touch with "the wildish nature," then they are better able to live "a vibrant life in the inner and outer world."

Although the image of the wolf may be more romantic, Tami is more comfortable with the metaphor of a bloodhound. She explained to her former employee—and now author and host of the *Emerging Women* podcast—Chantal Pierrat about how she enjoys "being able to be like a bloodhound smelling the next part of the trail, having all my senses intact, and not having my senses offline because I'm in a fear state."

Tami has a gift for finding the scent of genuinely gifted spiritual teachers. She started recording conferences featuring legendary leaders like Ram Dass and Stephen Levine and later Pema Chodron and Adyashanti, among many others. She worked diligently to improve her sound-editing skills and began building her business through catalogue sales. Tami's work became even more vibrant as she integrated her approach to sharing the treasures of the inner world with her evolving understanding of the commercial outer world.

In the quest to bring forth the "living wisdom" of the world's spiritual traditions, she realized that, rather than continuing to focus on programs recorded at conferences for large audiences, a studio setting could create a more intimate relationship with listeners, allowing the author's charisma and "energetic transmission" to be experienced by listeners. In 1990, Sounds True released a studio recording of *Women Who Run with the Wolves* by Clarissa Pinkola Estés. The audio program was wildly successful. Estés's Sounds True audio program was released before her book and helped it become a national bestseller, a breakthrough that brought a much bigger audience to the company.

Nearly thirty years later, Sounds True has brought more than 1,500 audio, video, music, and book titles to an audience of millions and has hosted many successful online seminars and events including their renowned Wake Up Festival. A two-time winner of the Inc. 500 award as one of the fastest-growing privately held businesses in America, the company is widely recognized as a pioneer in providing life-changing, practical tools that accelerate spiritual awakening and personal transformation.

Fortunately for the people who work at Sounds True, Tami has been able to create congruence between the noble truths espoused in her products and the actual running of the business. Long before the phrases "triple bottom line" or "people, planet, and profit" became part of the collective vocabulary, Tami had a deep knowledge that these principles would inform every aspect of her business. Guided by a vividly clear purpose from the beginning, it was a *fait accompli* that harmonious relationships with all stakeholders and planetary stewardship were nonnegotiable essentials. Even though she had no business background, she had common sense, so it was also clear that cash and profitability would provide the fuel for manifestation of her ideals. Many businesses work by relying primarily on "OPM"—other people's money—and trying to leverage that for gain. But blessed with the seed capital she needed to launch her company, Tami's first business insight was: *Don't spend money you don't have.*

And for many people in business, the ends justify the means. But the essence of the Sounds True business philosophy is that the means and ends can't be separated. She explains, "I believe that the means and the ends are the same—because ultimately all we really have are means. The means, the path itself, should always be life-giving."

Tami was committed to not just *disseminate* spiritual wisdom but to actually *apply* it to the culture of her organization. How her people work together had to be coherent and in keeping with the company's offerings.

When people questioned her idealism with comments along the lines of, "Be realistic. Look, business is business," Tami would respond incredulously, "What does that phrase even mean? I don't get that. It doesn't mean anything to me. It is ridiculous. It's absurd to try to separate people's well-being from the way we do business."

In 1990, MIT professor Peter Senge published the landmark book *The Fifth Discipline: The Art and Practice of the Learning Organization*. But Tami had already intuited the principles of a learning organization and was practicing them in her company. She created a workplace that supported the wholeness and development of all its stakeholders. She sees the workplace as an incubator of personal growth and psychological health: "It is only through engaging in the relational dynamics of the workplace that our narcissistic tendencies are put under the spotlight, our withholding is put under the spotlight, our lack of motivation, anything. All of this comes out in a team situation, and the team will show you what you can't see yourself." Tami has artfully nurtured a culture where people freely give and receive feedback about how they need to grow.

At Sounds True, learning is important, but loving is paramount. Whatever our life's purpose, it is our way of expressing our love in the world. Tami is devoted to helping all her people increase "the sense of real connection with each other in the workplace and with the people we serve through our businesses." She understands that love that is not expressed is like a check that never gets cashed; it doesn't do anything for anybody.

The commitment to nurturing a sense of connection and to creating a work environment that celebrates the sharing of love manifested, beginning in 1990, with an invitation for Sounds True employees to bring their four-legged friends to the office. The Sounds True website proclaims: "We are grateful for the cheerful and loving energy that comes from sharing our space with our canine companions—as many as fifteen on any given day!"

Tami emphasizes that a *learning organization* must also be a *loving organization* and that when those two elements come together we can realize the ideal of The Healing Organization. She explains, "Part of our healing is our willingness to look directly at the parts of ourselves that are in shadow and that are underdeveloped; the parts of our conditioning that we have yet to look at, embrace, and decide that we can transform and grow out of. This only happens when people feel loved and encouraged to learn."

The Sounds True culture supports people in being their authentic selves. As one employee expressed it: "Here, everybody has full permission to take their masks off; there is no posing. If somebody is feeling sad, they are free to express that. If they need to go home, they are free to do so; they do not need to make up any fake reasons."

Tami sees a strong link between such authenticity and healing. "This idea of being authentic at work and how that leads to authentic connection is, I think, a core idea of what creates healing. I don't know how much genuine anything can happen when people don't feel they can be genuine."

It's clear that the culture of Sounds True is a reflection of Tami's nature, her essential being. Tami says, "The one thing I've always known is that I would keep my soul intact. I have a thin line I can walk for my soul to be intact. If I veer off of it, things aren't good for me. I have to stay on it. When I do, my life unfolds beautifully. I think everybody kind of knows what they need for their souls to be healthy." The culture that Tami has crafted and leads at Sounds True is designed to create an environment that facilitates or makes it easier for people to be in alignment with their own souls' truth or expression, and their mission and purpose are focused on helping their customers do the same.

Although the organization mirrors her being in remarkable ways and many people look to her for inspiration, Tami doesn't like to present herself as a role model. When she was in her early twenties, she tried to find one for herself but it didn't work. She realized that what she was seeking wasn't to be found in someone else's story, but through what Emerson called "self-reliance." She thought: "I'm going to see what I am, and be true to that."

Over the decades this understanding has guided her and she believes it is an important message for anyone who is looking for a role model. She says, "I think a lot of times we are looking for role models to give us permission to be ourselves. There is no one else that's you anyway. No one else has your exact DNA, your parents, your exact combination of promises you made when you came here, to deliver to the world." She continues,

> If I could be a role model for anything, it is for this: *Fly your own freak flag*—without a role model. It could look really different for you and you have to find that and trust that, whether or not there is a seat already made for you at the table.

Of course, we all need as many examples of healthy expression and living life authentically as we can get, especially because we have so many egregious examples of the opposite. The danger is that an individual may invest too much in attempting to imitate another, just as many companies spend too much time and money benchmarking other companies. We must all find our own unique expression. Reading the stories of heroic individuals and studying the lives of healing leaders provides inspiration and offers clues we can all apply in finding our unique path.

Introduced in the 1970 anthem "Almost Cut My Hair" by folk-rock supergroup Crosby, Stills, and Nash, the phrase *let your freak flag fly* began as a motto celebrating the long hair of hippies, which, fifty years ago, seemed like a bold expression for those who wanted to discover and live their own truth. The Urban Dictionary notes that the phrase "implies unique, eccentric, creative, adventurous, or unconventional thinking."

Over a hundred years ago, Florence Butt was determined to discover and live her truth. As a woman who graduated from college—the only female in her class—and then started her own business, going door-to-door to get her enterprise going, well, you can bet that she was definitely considered eccentric and unconventional . . .

A year before the CSN song debuted, the Beatles began meditating with the long-haired Guru Maharishi Mahesh Yogi, which definitely seemed unique, eccentric, creative, adventurous, and unconventional at the time. Today, meditation and mindfulness training are part of the culture not only at tech companies like Apple, Google, and Intel but also at Aetna, Goldman Sachs, Procter & Gamble, and many others. Spiritual wisdom is being disseminated at previously unimaginable levels and Sounds True has made a significant contribution to this transformation of consciousness—because a college dropout pledged allegiance to her own freak flag.

BRINGING MORE PURA TO THE CORPORATE VIDA

HOW FIFCO TRANSFORMED FROM A POLLUTER TO A SAVIOR OF THE ECOSYSTEM

P*ura vida* or "pure life" is a phrase expressing the essence of the spirit of the beautiful country of Costa Rica. One quarter of its territory is in protected areas or national parks. Although the entire country is about the size of West Virginia—less than 0.04 percent of the planet's landmass—it is home to 5 percent of Earth's biodiversity.[1]

FIFCO, one of the world's great examples of a Healing Organization, is a company that has transformed so that it not only stopped polluting this natural treasure, but is now actively protecting it, while also showing other businesses around the world how to make more money by being better stewards of *Pachamama*, Mother Earth.

Incorporated in the little Costa Rican town of La Florida in 1908, The Florida Ice and Farm Company (FIFCO) started as an agricultural and ice manufacturing business. Over time, the company evolved to focus primarily on brewing beer.

How did this transformation happen? It wasn't a sudden enlightenment but rather a gradual process that began with Ramon Mendiola listening to his stakeholders, and eventually bridging the gap between business-as-usual and applying the values he learned from his parents.

Growing up near San José, Costa Rica's capital, he enjoyed a comfortable, though not ostentatious, upbringing. Ramon's father was a businessman who emphasized and modeled integrity and excellence. Ramon says, "My father taught us the importance of being responsible in everything that you do in your life, and always doing your very best." Ramon says that his mother has always been his conscience. "She reminded me of what is truly important in life: helping people around me, those who are not fortunate enough to have everything we have. All those things stayed with me since I was a kid."

After finishing college and getting his MBA at Babson College, Ramon worked at Kraft Foods and Philip Morris for eleven years. The companies were part of the same corporate entity, though Ramon worked on the tobacco side. During this time, he didn't really question the societal impact of the products he represented. He functioned as a "normal businessperson," focusing on building brands, capturing market share, and generating profits. He shares that his greatest motivation during this phase of his career, besides his deep desire to excel, was fear: fear of what activists or regulators might do to constrain his activities. During this time, he cultivated relationships with government officials and health and environmental groups that would prove to be creative collaborators when he moved to FIFCO.

At FIFCO, he was still motivated by the same forces that drove him in his previous job. Going from tobacco to alcohol and sugary soft drinks didn't lighten the pressure he felt from regulators and watchdogs of all kinds. He also felt the pressure of making sure that the hundred-year-old enterprise would endure, and for that to happen, changes would need to be made.

FIFCO's growth and profit were steady but not what they could be. Ramon was determined to accelerate the company's financial growth. His first two years, from 2004 to 2006, were focused on becoming more efficient. With the help of consultants, he applied best practices gleaned from global multinationals and his own business education to eliminate inefficiencies, reduce bureaucracy and complexity, and align compensation more closely with performance (defined then in purely financial terms). Ramon separated the more profitable beer business from the newer nonalcoholic drink division, bringing more accountability to each of the business units. Many people left the company during this time, uncomfortable with the new level of intensity and accountability.

The next stage, from 2006 to 2008, was to drive aggressive growth. He challenged his executives to double the size of the business in terms of revenues and profits within two years, with 70 percent organic growth and 30 percent new business expansion. Traditionally, the company had only grown 1 to 2 percent a year. With aggressive innovation in new flavors, pack sizes, distribution channels, new products such as iced tea, and new segments, the company was able to reach Ramon's goal of doubling in size by 2008.

But the company was starting to face a societal backlash to its aggressive growth. A lot of the growth came from the alcohol business, leading to widespread criticism as drunk driving and alcoholism rose. The company was also criticized for its sugary drinks, which were blamed in part for rising obesity. FIFCO was also criticized for its impacts on water, carbon emissions, and solid waste.

As Ramon recalls, "We had started receiving signals from society, especially around alcohol. When we started growing our business rapidly, we overheated a little bit in terms of the consumption of alcoholic products. We then had issues with harmful use of alcohol, including drinking and driving." Ramon started paying careful attention. He explains, "We were good at listening to our consumers and customers, but we decided to do a comprehensive qualitative and quantitative consultation with all our stakeholders: NGOs, government, society in general, those who regulate us, our shareholders and employees."

The feedback that Ramon got from stakeholders was clear. It identified four major negative footprints. The first was the perception that the company was promoting excessive consumption of alcohol. The second was about solid waste: people complained about FIFCO bottles floating in the rivers and washing up on the beaches. The third issue was water: FIFCO was seen as taking water away from communities to produce its products. The fourth was carbon emissions from the company's manufacturing facilities and distribution system.

This was Ramon's moment of truth, his opportunity to blend the qualities he learned from his father—integrity, responsibility, and the quest for excellence—with the lessons imparted by his mom—empathy, caring, and compassion. Ramon committed FIFCO to becoming a "triple bottom line" company focused on people, planet, and profits. He realized that social and environmental issues needed to be approached with the same rigor he had brought to growth and financial performance. This called for the company to integrate its business strategy with its environmental and corporate social responsibility initiatives.

The most pressing social issue was excessive alcohol consumption. Not knowing how to tackle the challenge, FIFCO partnered with an NGO in Canada that had been changing alcohol consumption patterns in Québec for twenty-five years, primarily through reducing binge drinking and encouraging moderation. FIFCO's research found that consumers drank on average less than two times a week, but had as many as five drinks on each of those occasions. The win-win outcome seemed clear: change consumption patterns by lowering the drinks per occasion while increasing frequency. As Ramon says, "Alcohol is not good or bad as a product. It depends on who consumes it and how: the consumption pattern." Working with the Ministry of Health and Education, FIFCO made significant strides in improving consumption patterns in Costa Rica and beyond. Ramon became the president of the Latin American Brewers Association and started to mobilize the industry to understand and shift consumption patterns in countries from Argentina to Mexico.

Alcohol consumption was moderated through these efforts, and incidences of drunk driving became less frequent. But Costa Rica was also confronting a dangerous increase of obesity, as were many other Latin American countries. Mexico has the highest rate of consumption of carbonated soft drinks in the world, and Costa Rica was not far behind. The Minister of Health and Education asked Ramon what his company could do to reduce the sugar content. Ramon committed to reducing the sugar content across his portfolio within three years. The only products he could not alter were the Pepsi brands that FIFCO produced under license. But he could work on the 70 percent that were not Pepsi related.

Ramon started by removing sugary carbonated beverages from vending machines and high school cafeterias, leaving only fruit juices and iced tea and similar products. He then put together a team, including his top R&D people, to work on reformulating all 186 stock keeping units (SKUs). They were asked to reduce the amount of sugar and replace it with natural low-calorie sugar substitutes. The team was asked to use artificial sweeteners only sparingly, with a goal to move toward all-natural (it is now 80 percent natural).

The team was able to reformulate all 186 SKUs within four months. Across the portfolio, sugar and calories were reduced by 40 to 50 percent. Ramon went back to the Minister of Health and Education with his technical formulations and three boxes containing the new products. He told her that, within a week, FIFCO would start producing the new formulations. She was stunned. "But, Ramon, you had asked me for three years to do this! How were you able to do it in four months?" Ramon replied, "Our R&D team did an amazing job."

The new formulations were launched without any public announcements, just modifications in the ingredient lists on the labels. The company monitored the market's reaction very closely. There were a few complaints about reduced sweetness with some of the SKUs, but for the most part, people did not notice. After two years, when consumers had completely embraced the products, the company launched an advertising campaign emphasizing how much healthier the products now were, and changed their labeling. Ramon was eventually able to motivate PepsiCo to let him reduce the sugar content in their products as well. As an example, a can of 7UP went from 150 calories down to 100 as part of a new global initiative by the multinational corporation.

The company also embarked on ambitious environmental steward-ship initiatives.

In 2008, FIFCO made three dramatic, public commitments for its environmental impacts: It would become a zero solid waste company by 2011, water neutral by 2012, and carbon neutral by 2017.

FIFCO accomplished all three of these goals using a simple three-step process: Measure the footprints precisely, reduce them to the maximum extent possible, and compensate externally for the difference. All of this has been certified by established, credible third parties. FIFCO has done this by investing in environmentally ad-vanced technologies, educating stakeholders, and linking environ-mental performance to compensation. The company has also invests 7.5 percent of its net profit every year to create projects to reduce footprints.

FIFCO reduced its water consumption from a peak of 10.9 liters used for every liter produced to 4.38. They compensate for the re-maining gap by providing clean water to people who do not have access, harvesting rainwater for schools and communities, and pro-tecting forest areas in the countries where they operate.

Reducing solid waste is accomplished by recycling, reusing, and generating energy from what is left. FIFCO has been able to get to 99.4 percent reduction, and compensates for the rest through a vol-unteer program that picks up trash generated by other companies. The company currently recycles 82 percent of all plastic bottles that it sells and 77 percent of all other packaging materials. Its goal is to recycle 100 percent of its bottles and also to replace them eventually with the use of biopolymers instead of PET in its bottles.

Most excitingly, the company is not content to get to zero negative impact. It is aiming at becoming a carbon, water, and solid waste *net positive* company. It wants to compensate at least 10 percent more water, fix 10 percent more carbon dioxide, and collect 10 percent more solid waste than it produces by 2020. It is working with its entire value chain to realize these goals. It started with its most prominent beer brand, Imperial, which became the world's first water positive beer in 2017.

What do we learn from FIFCO's experience in not only mitigating but in many ways reversing its environmental impact? Here are six lessons:

1. Listening to and honoring all stakeholders to better understand the problem as well as participate in the solution.
2. Having a higher purpose that encompasses flourishing in all dimensions.
3. Taking a holistic approach to business, which looks at social, environmental, and financial performance simultaneously rather than sequentially.
4. Linking measurements and compensation to holistic goals. Forty percent of compensation for FIFCO leaders is linked to environmental and social goals, likely headed to 50 percent in the future.
5. Maintaining a visible, public commitment to tangible goals. This creates pride, enthusiasm, and optimism among employees and other stakeholders.
6. Collaborating widely. None of these challenges can be solved alone, especially the more significant ones. You need to partner with academia, civil society, and the government—not just locally but in many cases globally.

When he started listening to stakeholders' concerns around the company's footprints, Ramon's consciousness was still very much rooted in fear. But after he started engaging with them, he began to see that the company could have a tremendous positive impact. Fear dissipated and was replaced by excitement and then joy.

Ramon made a series of bold public social commitments, having to do with water, carbon, and solid waste. The company delivered on all of them on schedule, leading to a surge of pride in its employees. This was a great insight for Ramon: doing the right things makes people happy.

Even though the company's environmental and social initiatives were all externally directed, Ramon found that the biggest impact was on his people: "The engagement and commitment of my most important stakeholders, which are my employees, went up tremendously." That creates a virtuous cycle: doing the right things leads to greater employee commitment and engagement, which leads the company to be even more effective at doing the right things.

•   •   •

In 2014, Ramon could look back on a decade of extraordinary progress. He had set ambitious goals and realized nearly every one of them. But he wasn't satisfied. "Although we were doing lots of positive things there was something missing, a clearly defined, shared sense of our higher purpose." Working collaboratively with a range of stakeholders, FIFCO defined its purpose in a simple sentence:

We bring a better way of living to the world.

Perhaps these words sound innocuous or trite, but for Ramon and his team they carry great power and meaning. They elegantly express the spirit of the company's focus on creating opportunity, well-being, and growth for all of its employees, providing delightful and healthier products to customers, while generating robust financial returns for investors. And doing all of this in a way that avoids harm and actually improves the environment.

Ramon traveled to all of the company's locations to share the purpose directly with his people. At one of these meetings, an employee took the microphone and said, "Ramon, I am very proud of everything this company has been doing in terms of water, emissions, our policy on alcohol, all that. I read about it, I see you on TV, it's all great; but, Ramon, do you know that we have employees living under poverty conditions?"

Ramon was stunned, and humbled. He paused and then replied, "Thank you for pointing that out. I don't know how many of our employees are living under poverty conditions, but allow me six months to work out a comprehensive plan with my management team."

Ramon and his team embarked on a project to understand the problem and devise a plan for how to address it. FIFCO hired a team of social workers to help. Ramon went with them and interviewed the people living in poverty: not just employees but also their families. He wanted to understand how and why these people were impoverished. He knew that it was not because of inadequate compensation; FIFCO pays some of the best salaries in Central America.

Approximately 18.5 percent of Costa Ricans live in poverty as reported to the United Nations, and 4.2 percent live in extreme poverty.[2] As a relatively prosperous country in Central America, Costa Rica has experienced a disturbing phenomenon: Many people escape poverty only to later regress. FIFCO's research revealed that one of the primary causes was the vicious cycle of temptation and debt that begins with the aggressive and seductive marketing of luxury goods. In the capital city of San Jose, store windows at electronic outlets are plastered with large signs advertising the low monthly installment cost of, for example, a large flat-screen television. Similar signs are prominent in the showrooms of dealers selling fancy scooters and motorcycles. But when you study the fine print, you'll find the startling disclosure that the underlying interest rate ranges from 50 to 60 percent a year.

Financial literacy is a problem everywhere, but especially in poorer countries. Aspiring to a better life, people give in to the temptation of "low monthly payments" to acquire products they can't afford. They end up paying a huge amount over the actual cost of the purchase—if they are able to keep up with the payments. If something happens in their life that causes them to miss a payment, they accrue additional charges, which they have no hope of ever being able to pay. Soon, they owe far more on the item than what it is worth. The items get repossessed and sold to others, while the crushing debt remains.

Additional factors that cause employees to be in a vulnerable financial situation include making other bad financial decisions, health issues, and having too many dependents. Overall, about 3.6 percent of FIFCO's Costa Rican employees were living in poverty conditions: 161 employees and 644 relatives who were dependent on them.[3]

At the end of six months of study, the company launched its program: FIFCO Oportunidades (Opportunities). Ramon made another bold public commitment: While it takes the government ten years on average to bring a person out of poverty, FIFCO would bring every single employee out in three years or less. With the input and agreement of the families affected, the company developed a customized initiative that included providing a mentor for each of the families to help them learn practical financial literacy. One of the toughest cases was assigned to the company's CFO. Ramon told him, "If you cannot help this family get out of poverty, nobody can!"

In their research, the social workers also encountered a lot of men's issues, what Costa Ricans call "machismo." So the program included supporting men around how to be better husbands and better fathers. Overall, the program included education and support in four areas essential for human development and well-being: nutrition and healthy lifestyle, housing, education, and comprehensive family financial management.

FIFCO partnered with banks and government ministries to help employees refinance high-interest loans, get assistance for housing, or for health issues. The goal was not only to rescue people from their immediate financial exigencies, but to teach them how to prevent those from happening again in the future.

As promised, by December 2018, three years after the program was launched, 100 percent of the group were officially out of poverty. In 2018, FIFCO expanded the program to its Guatemalan employees. It plans to extend the program to the US in 2020.

The program takes an investment of approximately half a million dollars a year from FIFCO. Over three years, this came to $1,875 for each of the eight hundred people impacted. Ramon says, "This is a small price to pay for the impact on all of those lives for the foreseeable future. It could even have a multigenerational impact as children grow up in more secure households and learn about financial and other kinds of responsibility at a young age."

So what has been the impact of this program on the company? Beyond the obvious impact on the 3.6 percent, there is a more subtle impact on the remaining 96.4 percent. As Ramon says, "The employees are very grateful. They see our company really thinking and doing something about this problem and taking care of their colleagues." FIFCO's rating as a great place to work was already high, but it has skyrocketed even further as a result of this program.

While the program has been an undeniable success, Ramon is saddened by the fact that other companies seem reluctant to follow suit. The habitual business mentality is, "Let's fire these irresponsible people and hire new ones who won't get into this mess to begin with." Most companies in most countries think that alleviating poverty is the government's responsibility, or that it is up to nonprofits. Ramon challenges this assumption: "What if, instead of looking to governments to solve poverty or blaming them, we start looking inside first, inside our own organizations?"

Getting people out of poverty was crucial, but it is also important to keep others from falling into poverty. At FIFCO, another 2 percent of employees are on the brink of poverty. So the company expanded elements of its program to include that group. Resources are available and proactive interventions are in place to make sure that no employees in the future fall into this trap.

Having successfully completed the initial three-year program, Ramon is now sharing the experience widely with the outside world.

FIFCO's ultimate goal is "to inspire other companies around the world to join this initiative and make a pledge to eradicate poverty within their employee base. Poverty can be eradicated if we do it in one organization at a time. The private sector has a major challenge: that of complementing the government's efforts and working to construct a fair and more equitable society."

• • •

FIFCO complements its Oportunidades initiative with a remarkable employee volunteer program. Called Elegí Ayudar, or "choose to help," the company contributes two working days of community service work for each of its 6,300 employees, approximately 70,000 hours a year. By 2016, the cumulative time donated was 450,000 hours. Ramon wanted to take it further and made another public commitment: by 2020, FIFCO will have completed a million hours of volunteer service.

A cross-functional team working on attaining this stretch goal came up with a brilliant idea: Engage *all* the stakeholders of the company in reducing and reversing the company's environmental footprints, starting with the families of employees. So the company started by inviting families with children fifteen or older (for insurance and liability reasons) to engage in meaningful volunteer work, which they loved. It became a way for families to come closer together and to teach family members about real issues facing communities and the environment.

The next element was even more significant: engaging with consumers. They started with a couple of pilots with their two most important brands. These pilots proved tremendously successful in engaging consumers in helping reduce the company's footprints, generating a huge amount of publicity and social media postings. In 2018, FIFCO created more than twenty experiences with consumers, with all of its major brands participating. Ramon is delighted. "Consumers nowadays, especially millennials, want to transcend, they want to be part of something, they want to have meaning."

FIFCO has started to include other stakeholders as well in volunteer efforts, including shareholders, suppliers, clients, and other partners. The company handles all the logistics: Volunteers come in at eight in the morning, the company provides transportation, safety equipment, and meals. Consumers, shareholders, suppliers, and employees work side by side, strengthening relationships, generating goodwill, and making an actual difference in the world.

The volunteer program is now called "Choose to Help 4.0" and it offers all stakeholders an emotionally charged experience of connection to each other with a higher purpose.

Ramon continues to set ambitious goals, challenging his people to accomplish the seemingly impossible. So far, he has succeeded every time, and in the process has empowered people to become agents of healing. He has gone from reducing suffering to healing and bringing joy, from not doing harm to actually doing good, moving from negative to positive impacts in every dimension.

• • •

When Ramon Mendiola joined the company as a board member and then as CEO in 2004, it was a traditional, established company with about $150 million in revenues, 1,800 employees, and a negative environmental footprint.[4] Fast-forward to 2018. Revenues are over $1.2 billion, and the company is thriving with over 6,500 employees and more than $260 million in annual profits. Its businesses now span beer, spirits, a wide range of nonalcoholic beverages, as well as hotels and retail operations in Central America and in the US. Recognized as the best place to work in Costa Rica, it is one of sixteen companies formally certified as "sustainability champions" by the World Economic Forum.[5] FIFCO is among the 1 percent of companies in the world that allocate more than 7 percent of profits toward social investments.

Ramon reflects, "This decade has been life-changing for me. I have redefined my success as a CEO and I have gotten closer to my executive team and all my employees in a more meaningful way. Volunteering with them, sharing the passion and excitement of accomplishing social and environmental endeavors, we have grown together. I have realized what my own purpose in life is now: using all my energy to motivate other business leaders to take this path."

Ramon's accomplishments are being recognized and he's increasingly invited to participate in major forums around the world with leaders from business, government, and NGOs. "I have the honor to motivate them and demonstrate with concrete evidence that a company can double or triple its economic value while at the same time becoming a positive force for the environment and society."

Ramon is a practical visionary who has given the national motto of his country deeper meaning: the *vida*, the life of all his stakeholders, has become more *pura*, more pure in every dimension as a result of his efforts. "We can no longer operate under the premise that businesses can extract the earth's resources with just the responsibility of generating profit, employment, and paying taxes." He adds, "We have so many difficult social and environmental issues and our governments do not have the resources or competence to deal with them. We, the private sector, are now the world's main hope for finding solutions."

CEO = CHIEF EMPATHY OFFICER

HOW HYATT'S CULTURE OF CARING, MINDFULNESS, AND WELL-BEING OFFERS A MODEL FOR ALL COMPANIES

I n 1881, Nicholas Pritzker and his family fled persecution by the Tsarist Russian Empire and came to the United States, where they settled in Chicago.[1] Nicholas pursued the American Dream through education and hard work, working first as a pharmacist and then becoming an attorney and entrepreneur after graduating from Northwestern University School of Law. He wrote a business manual that he passed on to his sons, emphasizing that the key to immortality was making a better world for future generations. Pritzker's progeny became successful entrepreneurs in a range of businesses and generous philanthropists as well. In 1957, grandson Jay purchased the Hyatt House motel adjacent to the Los Angeles International Airport—the first holding of what was to become one of the world's great hotel enterprises.[2] Commenting on the family business approach to acquiring companies, Jay Pritzker explained, "Our philosophy is that we don't buy companies to strip them of their assets. The key to us is the people who run them."

Today, Hyatt Hotels Corporation is a multibillion-dollar global company employing over 110,000 people in 56 countries who speak over 100 languages. In 2006, Hyatt asked Mark Hoplamazian, who had served the family in a number of roles since 1989, to become CEO, with a mandate to grow the company by strengthening its caring culture.

Mark's parents were first-generation Americans of Armenian descent. The Pritzkers were Jews who fled the Russian Empire and the Hoplamazians were Christians fleeing the Ottoman Empire, both looking to find freedom and opportunity in the United States of America.

Mark's mother endowed him with a spirit of gratitude and with a love of learning that led him to graduate from Harvard University and then complete an MBA from the University of Chicago. His parents also raised him with sensitivity to the suffering of fellow humans and a sense of responsibility and care for others.

Mark came into his role at Hyatt with no background in the hospitality business. Blessed with a beginner's mind, he was not afraid to ask innocent questions. He explains, "My ignorance was a powerful asset because it led me to ask a whole bunch of naive, simple questions that led to discussions about why we do certain things the way we do, and that led to many positive changes."[3]

Mark led his team to contemplate fundamental existential questions beginning with:

What makes Hyatt unique?

That led the team to think deeply about Hyatt's history and why they joined and why they stay. As he met with hotel teams from around the world, many of whom had been with Hyatt for a long time, a consensus emerged: *caring* was the key distinguisher of their culture.

So the next big question was:

How do we nurture our culture of caring and scale it so it thrives in all of our global locations?

Given that this was already a part of Hyatt's DNA and its sixty-plus-year history, it was a matter of consciously acknowledging it and making it more explicit and tangible. The leadership team engaged in a thorough exploration of the factors that promote genuine caring, as well as those that may inhibit it. The team recognized that caring begins with understanding and empathizing, but that empathizing with someone does not translate into care *unless you do something about it*. They summed up this insight in a simple formula: E + A = C: *empathy plus action equals care.*

And they clarified their purpose with the following declaration:

We care for people so they can be their best.

Those ten words carry a lot of meaning within Hyatt.

Much of the focus in the hotel business is on service, but *care* is different than service. It is possible to care for someone without serving them. But it is *definitely* possible, and quite common in many so-called service businesses, to serve people without caring for them. Hyatt seeks to bring service and caring together.

The second key word is "people." The Hyatt purpose is not just about guests, it's also about colleagues, suppliers, members of the community, shareholders—all of the company's stakeholders. Like all Healing Organizations, Hyatt realizes that when employees experience caring and support, they do a much better job and the result is happier guests and wealthier shareholders.

Lastly, the phrase "to be their best" also has a precise meaning for Hyatt. As Mark says,

We're not in the business of making people feel a magical state of excitement the way Disney does. Guests are with us for various reasons: weddings, funerals, celebrations, vacations, seminars, award ceremonies, product launches, and many other activities. We experience real life with real human beings every day, and so helping someone be their best is not helping someone achieve one specific outcome. Rather it's helping them so they can actually be at their best, whether they're with us for a business meeting or a vacation.

For Mark, this wasn't just business; it was personal.

His deep commitment to the core value of caring as an expression of empathy plus action led him to employ a range of assessment tools to help the leadership team understand their own strengths and weaknesses. One of the dimensions that the tools sought to uncover was empathy. In looking at his own results, Mark was shocked to find that his score was low relative to others at Hyatt. He thought of himself as a paragon of empathy, and his initial instinct was to reject the data. But as he reflected, he realized he needed to check in with the people who know him best: his wife and his children.

When he became CEO, Mark arranged his schedule so he could drive his kids to school every day and engage with them and cultivate their relationship. During one of the drives, he shared the feedback he had received from the assessment and was shocked when they confirmed that he wasn't as good at empathy as he thought he was. They explained, "Well, most of the time you're with us you're doing something else . . . you're on the phone or responding to emails. In the car we have to tell you to look up because the light turned green and you're supposed to go." Mark's wife gently but firmly confirmed the children's feedback.

Although initially disturbing, this led Mark to an epiphany. He realized that empathy requires humility, intentionality, and presence. Mark explains, "In order to practice empathy, you have to be present, and one great vehicle to being present is to be mindful."

This realization led to Hyatt cultivating mindfulness as a central part of the company's commitment to caring and well-being. This investment is more than just philosophical: Hyatt recently bought Miraval, a destination spa resort company, and Exhale, a fitness and spa company. The programming and the ethos of both acquisitions are based on mindfulness in nutrition, movement, self-care, and caring for others.

Hyatt is leveraging these investments to better fulfill its purpose, because helping people to a greater state of well-being is a clear way of caring for them.

Hyatt has crystallized its efforts in this area by focusing on three aspects of well-being:

1. *Feel:* How you feel, or your emotional and mental well-being.
2. *Fuel:* How you fuel your body through things like food, nutrition, and sleep.
3. *Function:* How you physically move and function at work, in life, and at play.[4]

Hyatt is making tangible investments in each of these areas as well as entering into strategic collaborations with partners to elevate how all the stakeholders feel, fuel, and function when they are at a Hyatt hotel (as well as in the rest of their lives).

An important aspect of this is Hyatt's long-standing global food philosophy, expressed as "Food. Thoughtfully sourced. Carefully served." Hyatt's commitment is to offer food options that are good for guests, the community, and the planet.

When you have 110,000 employees and they enjoy a healthier diet, when they're encouraged to take advantage of free fitness facilities, and when they're offered the opportunity to learn meditation, those employees feel cared for and they share that caring with guests and other stakeholders. Moreover, days away from work due to illness and other health-related costs drop significantly.

Promoting well-being alleviates suffering on many levels. Mark knows that many of the guests who are staying at one of his properties work for organizations that aren't healing, but hurting. The pace of change and massive disruption across many different industries has caused many people to experience intense pressure. Stress-related illnesses are on the rise. As a society, we don't do much to facilitate wholeness and well-being; we are focused on diagnosis and recovery from illness. As Mark told us, "The health-care system is really an illness-care system, and we don't really focus our time and attention on true well-being and the means by which we can actually help promote wellness within the workplace." Mark intends that staying at a Hyatt will help inspire guests to greater well-being not just during their stay but after they depart.

Gratitude, caring, and love are the messages that are consistently reinforced through Hyatt's public communications. For example, the company aired an advertisement on the 2017 Oscars broadcast set to the soundtrack of Burt Bacharach's classic song "What the World Needs Now Is Love, Sweet Love" to emphasize the need for understanding of others in the world today. It was more than just an advertising campaign. The company has taken a leadership position in environmental and social issues, supporting gay rights and the Dreamers immigration issue, both of which have strong resonance with its values as well as the concerns of its employees and guests. As Mark says, "I have a responsibility as a CEO to speak up on issues that are relevant to our colleagues and to our guests, because it's partly how we define what the brand stands for. People make a choice to affiliate with the brand, either to come work for us or to stay at one of our hotels. They need to understand that this isn't just a public relations initiative, we're serious, we stand for caring, for all people."[5]

Hyatt touches hundreds of thousands of employees and guests every day; millions of people come into contact with the company every year. Mark's passion and love shone through in our interviews with him and he shared many stories of how the culture of caring comes to life on a daily basis at Hyatts around the world. So much so that we call him CEO: Chief Empathy Officer. Here's one that's typical:

A guest at a New York property called the front desk because of a problem with the air-conditioning in her room.[6] As he was repairing the air conditioner, the engineer heard the woman talking on the phone and it soon became clear to him that she had just learned through that phone call that her mother had passed away. He could've pretended not to have noticed and just gone about his business. Instead, he bore witness to her suffering and when she hung up the phone, he offered kind words and his caring presence. The guest was deeply moved by his kindness and later wrote a note to him and the company about what he had done and how much it comforted her and touched her heart that she felt his caring in that very painful moment. A year later, she wrote again sharing her gratitude and amazement that the engineer had taken the trouble to write a follow-up note to her to check in and see how she was doing.

Mark says, "If we can actually have an impact on the lives of the people we touch and give them a better path for their own well-being or to help them along in their process of healing, then we've done something wonderful."

Mark clarifies, "Our vision is clear and simple: to achieve a state where we've elevated the level of understanding and care in the world." The descendants of immigrants who fled empires to pursue the dream of freedom and prosperity came together to create a ministry of caring and well-being. That's the real American Dream and a vision for our world. That's the vision of a Healing Organization.

PART 3

THE BECKONING PATH: BECOMING A HEALING ORGANIZATION

THREE PRINCIPLES
THAT DEFINE
A HEALING
ORGANIZATION

We have offered a broad range of examples of Healing Organizations, spanning diverse industries such as carpet weaving, hospitality, fashion, energy, industrial machinery, grocery, publishing, consulting, and construction, ranging in size from a few hundred people to well over a hundred thousand. Some serve just one city, state, or region while others are global in scope. Some got started recently, and others have been around for more than a century. Some started from their first day with an intentional devotion to meeting people's real needs, alleviating their suffering, and elevating joy in their lives. Others started principally as moneymaking enterprises, but later discovered the power, beauty, and joy of becoming catalysts for healing in the lives of all they touch.

Whatever the industry, geography, duration, or process, Healing Organizations share certain qualities and fundamental beliefs. They all operate with a deeper understanding of what it means to be "in business," based on a deeper understanding of what it means to be human. They refuse to add to the ugliness in the world; instead they reflect the deepest aspects of what it means to be human, which is to embody and manifest the Platonic ideals of the Good, the True, and the Beautiful. They recognize and celebrate goodness, they have an unshakable commitment to the truth, and they treat people beautifully. And they all discover that operating in this way yields more abundance.

Different organizations formulate and articulate these principles in different ways, but ultimately they all share core values and principles that we will review and summarize here in part 3.

The most fundamental of these was articulated beautifully by Herb Kelleher:

The business of business is people—yesterday, today, and forever.[1]

This is an exquisitely simple and powerful statement. Everything else, including profits, should be viewed as a means to enable the flourishing of people and the planet.

Bob Chapman expressed a similar notion in different words:

We measure success by the way we touch the lives of people.

Herb and Bob both realized that when we define success as achieving power, money, and position for ourselves, we almost always inflict significant and unnecessary suffering on others.

Healing Organizations are people-centric, "Truly Human" to use Bob Chapman's term. In other words, they *put the human case ahead of the numbers*. They don't sacrifice higher values for lower ones. For Healing Organizations, fealty to values is foremost and success metrics are focused on fulfillment measures for all stakeholders in the long term. Profitability is important but it is viewed more as an epiphenomenon rather than a *raison d'etre*.

We have identified three core principles that distinguish Healing Organizations. They are:

1. Assume the moral responsibility to prevent and alleviate unnecessary suffering.
2. Recognize that employees are your first stakeholders.
3. Define, communicate, and live by a Healing Purpose.

ASSUME THE MORAL RESPONSIBILITY TO PREVENT AND ALLEVIATE UNNECESSARY SUFFERING.

Pioneering researcher into unleashing human potential F. M. Alexander wrote: "In order to do the right thing you must first stop doing the wrong thing."[2] The first principle for empowering healing is to become aware of the ways in which your enterprise may be causing or exacerbating unnecessary suffering and *stop*.

Our prime purpose in this life, according to the Dalai Lama, is to help others. His Holiness adds, "If you can't help them, at least don't hurt them."[3]

Primum non nocere is Latin for a pledge adopted by many physicians meaning *first, do no harm*.[4] Commit to causing no unnecessary suffering through your business. And, to the extent that you possibly can, avoid investing in companies that exploit workers, harm customers, weaken communities, and desecrate the earth.

The television show *Undercover Boss* features one episode after another in which leaders are shocked to discover the suffering that is going on in their own organizations. Don't wait for a camera crew to show up to seek out unnecessary suffering and alleviate it.

In order to translate this principle into real healing, we need to be open to discovering the gaps between our positive aspirations and the current situation. As Fred Kofman wrote in *Conscious Business*, "There are no death camps in corporations, but many apparently successful companies hide great suffering in their basements."[5] Healing Organizations turn on the light in the basement; they realize that just because you don't see it doesn't mean suffering doesn't exist.

Many people have a hard time asking for or accepting help. They don't want to appear unprofessional or needy, so they deal silently, often heroically, with crushing burdens.

This stoicism is an admirable quality, associated with courage and grit, necessary elements in life. But sometimes this attitude prevents people from asking for help when they really need it. Focus on searching for real suffering and cultivating an environment where people feel comfortable going to their team for support.

John Ratliff of Appletree Answers was shocked and humbled to discover that one of his full-time employees was homeless. His conscience was awakened and he began to, as Bernie Glassman, founder of Greyston Bakery, expressed it, *bear witness*. And as Glassman taught, when we are fully present and compassionate, "We don't have to figure out solutions ahead of time . . . loving action arises."[6] Ratliff's loving actions—including creating an in-house Make-a-Wish program for his employees in need that made a profound difference in their lives—arose naturally as he became aware and open to the challenges his people faced.

The corollary to *Primum non nocere* is *Malus Eradicare*, Latin for *root out evil*. A few simple ways to do this are:

Eliminate Hurtful Rules and Practices

Rigid rules, micromanagement, and draconian employee monitoring systems are often the cause of needless suffering. When Bob Chapman acquires a company, one of his first initiatives is to eliminate degrading and dehumanizing practices like body searching employees at the end of the workday to stop them from stealing. How can there be an atmosphere of trust if policies and procedures assume that employees can't be trusted?

Herb Kelleher emphasized, "If you create an environment where the people truly participate, you don't need control. They know what needs to be done and they do it."[7] Avoid policies that punish the 99 percent of people who are not opportunists in order to prevent the few who are from taking advantage. For example, The Motley Fool has no policy for sick days. As Tom Gardner told us, he doesn't worry about people abusing the non-policy. He explains, "We trust the Fools we hire; they're adults. The real problem in a purpose-driven company is that people will want to come in when they are sick. We don't want that. If you're sick, please stay home. We also don't have a vacation policy. We aspire to a culture of mutual trust. We focus on performance and let our people manage their own time."

Healing Organizations aim to replace rules with shared values. Richard Sheridan of Menlo Innovations explains, "We bring a positive assumption about our people to work every day. We assume noble intent and trust that everybody is doing what is in the best interest of the firm, their family, and the community."

Minimize the Hierarchy and Don't Incentivize Bad Behavior

Excessive hierarchy becomes oppressive and inhibits innovation and creativity. It gives those higher up in the system excessive power over those further down. This often has an effect similar to the Stanford Prison Experiment conducted by Zimbardo.

As Stanford professor Bob Sutton explains, "A huge body of research—hundreds of studies—shows that when people are put in positions of power, they start talking more, taking what they want for themselves, ignoring what other people say or want, ignoring how less-powerful people react to their behavior, acting more rudely, and generally treating any situation or person as a means for satisfying their own needs—and that being put in positions of power blinds them to the fact that they are acting like jerks."[8]

If your organization's incentive and compensation system rewards people exclusively for individual financial performance without reference to their cultural impact, you will likely be reinforcing hurting behavior.

Like Menlo Innovations, which transformed the model of software creation and reward from reliance on "towers of knowledge" to a rotating team approach, all of the companies we studied put tremendous continuing effort into structuring themselves in ways that promote collaboration and teamwork. They all seek to find innovative approaches to be sure that everyone, at all levels, shares in the company's financial success. This isn't easy and requires continuous monitoring and creative thinking. Be vigilant to ensure that you don't incentivize people to behave in ways that are not healing.

Never Enable Abuse

The work of Jane Dutton, Monica Worline, Bob Sutton, and many others has demonstrated clearly that one sociopath in a leadership position can throw off the culture of an entire company. *To spread healing leadership throughout the organization, never put up with sociopaths.* As Bob Sutton explains, "As much as I believe in tolerance and fairness, I have never lost a wink of sleep about being unapologetically intolerant of anyone who refuses to show respect for those around them."[9] And Herb Kelleher adds, "I forgive all personal weaknesses except egomania and pretension."[10]

This means parting ways with unrepentant toxic "leaders" who believe that generating big numbers makes it acceptable to denigrate others. At times, this may even require restructuring one's whole approach to how business is done. When Nand Kishore Chaudhary created Jaipur Rugs, he saw that women were being maltreated and abused. He realized that the whole system had to change. Against great resistance from other members of his caste, he banished the abusers from his enterprise and transformed the lives of 40,000 women and their families.

Don't Settle For Neutrality: Aim to Have a Positive Impact in All Dimensions

Google's original code of conduct featured a motto that became iconic, and in the case of many organizations, would still represent a great leap forward: "Don't be evil." Originated in 2001, this was an expression of the company's attempt to promote ethical behavior without sounding overly corporate. The official language of Google's code was recently changed and the new motto is: "Do the right thing." Of course, it may not sound as hip but it's better to think affirmatively. In the language of Steven Covey's classic habits of effectiveness: *Be Proactive.*

FIFCO, for example, became aware of the negative effects of their production process on the beautiful ecosystem of Costa Rica. At first they focused on just remediating the harm they were causing, but then they realized that they could actually go from negative to neutral to positive with all footprints, such as water, carbon, packaging waste, and health.

RECOGNIZE THAT EMPLOYEES ARE YOUR FIRST STAKEHOLDERS

In 1909, Harry G. Selfridge, the founder of the eponymous department store in London, introduced a revolutionary notion: "The customer is always right."[11] At a time when *caveat emptor* (buyer beware) was the dominant mindset in retail transactions, this gave Selfridge an advantage in the market and the slogan was soon adopted by many other businesses.

Since then, we've learned a lot about inspiring and delivering exceptional customer service. It turns out, as Danny Meyer discovered, assuming they are always "right" is not always the most intelligent or helpful way to consider issues with customers. Danny emphasizes that customer concerns are significant priorities but that employees must feel that they are supported by the company when a conflict arises.

Southwest Airlines has a remarkably responsive customer service orientation: they listen carefully to all complaints and aim to generate satisfaction. But one regular flyer was pushing them to the limit with a barrage of letters decrying every aspect of their operation from the no-assigned-seating policy to the type of peanuts that were being served. Exasperated, the customer service team forwarded the file to

Herb Kelleher, who immediately wrote her a simple note:

> Dear Mrs. Crabapple,
> We will miss you.
> Love, Herb.

By focusing on employees first, Healing Organizations generate exceptional customer care. They know that when employees feel cared for, they provide better service, especially knowing that they will be supported in dealing with customers who are difficult or abusive.

So create a culture of people looking out for and helping each other. Awaken a genuine sense of altruism within the organization: people doing things for others without expectation of something in return. Encourage creative expressions of caring; tap into the infinite creativity that exists in your stakeholders to stimulate the discovery of more ways to reduce suffering and increase joy. Give people a sense of psychological safety, so they can take reasonable risks and be true to who they are.

Many companies now practice customer advocacy. For example, if they believe that the customer's interests would be best served through a competitor's offering, they steer the customer to that offering. We also need to practice employee advocacy: having people within the organization look out for the best interests of employees.

A few of the best employee advocacy practices that we discovered in our study of Healing Organizations are:

Include Families as Stakeholders

Healing begins at home, in the lives of employees and their families. Consider employees' spouses, children, and, if applicable, aging parents who may need care, as stakeholders. At Sounds True and Menlo Innovations, even pets are considered stakeholders. Seek feedback from all these stakeholders (Well, we know what the pets will say: "More treats!"). In a Healing Organization, "work" and "life" aren't separate, competing entities that need to be "balanced." They can be integrated into a seamless whole. When people have a sense of purpose, meaning, and care in their work, and when the organizations in which they work prioritize their well-being and self-expression, work becomes one of life's noblest and most joyful experiences.

Look For Every Opportunity to Bring Joy, Play, and Love to the Workplace

Richard Sheridan didn't hesitate to say, "Bring your baby to work," when one of his people lost her childcare resource. Now, people visit from all over the world to find out how Menlo Innovations is able to turn out great software, on time, with exceptionally low turnover and superior profitability. They're able to do it because they've created a culture that "encourages people to bring their whole selves to work and apply the full range of their potential, energy, and talent."

Foster Healthy Internal Competition

The story of Cain and Abel set the tragic stage for sibling rivalry in many families and for unhealthy competition in many organizations. Instead, inspire "sibling revelry" as USHG founder Danny Meyer does, to channel natural competitiveness in a healthy way. Reject approaches such as "up or out" or "rank and yank"—zero-sum ways of being in which only a few can succeed and others must fail.

Invest in Healing the Way You Might Invest in Marketing or R&D

Many healing initiatives are not free. They require an investment, but the returns are likely to be far greater than almost anything else you can do. Be prepared to allocate resources to things that really matter and make a big difference.

Enable People to Continually Grow and Evolve

Help people evolve and overcome dysfunctional patterns. Resolve to leave people better than you found them, in all dimensions: mental, physical, emotional. Provide them continual opportunities to grow and evolve as human beings, not just as employees. All the learning experiences the Barry-Wehmiller Leadership Institute offers are designed to help people deal with the most important personal challenges they face—which also helps them immensely at work.

DEFINE, COMMUNICATE, AND LIVE BY A HEALING PURPOSE

What is the purpose of business? It is not, as Peter Drucker wrote, *"to create and keep a customer,"* and it is not, as Milton Friedman proclaimed, to make a profit. Creating and keeping customers and making profit are important measurements of success that are often mistaken for purpose. The purpose of business, now more than ever, must be to alleviate suffering and elevate joy by serving the needs of all stakeholders, including employees, customers, communities, and the environment.

Healing Organizations create and keep customers and generate profit in order to continue to grow and bring healing to more of the world. They are organized around a clear sense of purpose that shares some common characteristics. We express several of these in the acronym HEALING: Heroic, Evolving, Actionable, Loving, Inspiring, Natural, and Grounded.

Heroic

"The time has come for us all to help create and enjoy a new 'psychology of liberation,'" writes Dr. Philip Zimbardo in an essay, "Why the

World Needs Heroes."[12] For Zimbardo, a hero helps to improve the human condition "through acts of kindness, generosity of spirit, and a vision that always seeks to make others feel special, worthwhile, understood, and embraced as our kin." Craft a purpose for your life and your enterprise that will have a positive transformational impact on the world, impacting not only the company's stakeholders but also its industry and perhaps even society at large. Anything worth doing is heroic; it will stretch the organization and require it to undertake the seemingly impossible.

Evolving

A healing business aligns its purpose with the evolutionary impulses of its times. As we progress on our journey toward awakened conscience and consciousness, companies will have to adapt and elevate their purposes to remain in harmony with our aspirations and motivations. Eileen Fisher, for example, serves and supports the evolving expression of positive feminine energy, and Hyatt aims to bridge the gap between our current health-care crisis and an evolving understanding that we can make a huge difference in our own health by cultivating habits that support well-being.

Actionable

Craft a purpose that can be translated into action, so that "What" you do every day is directly linked to your understanding of "Why" you do it. When people relate their everyday actions to a higher purpose, miracles can happen. When DTE Energy defined its purpose statement—"We serve with our energy, the lifeblood of communities, and the engine of progress"—and shared it with the people at their power plants, employees were moved to tears. CEO Gerry Anderson recounts, "I learned an incredible lesson in leadership. I discovered what people are capable of when they really believe in something."

Loving

A purpose organized around love and care creates a powerful, vital force field throughout an organization. It is in harmony with the deepest essence of what it means to be human. This Bible excerpt from 1 Corinthians (13:4–8 NIV) expresses this element of a Healing Purpose beautifully:

> Love is patient, love is kind. It does not envy, it does not boast, it is not proud. It does not dishonor others, it is not self-seeking, it is not easily angered. . . . It always protects, always trusts, always hopes, always perseveres. Love never fails.

Herb Kelleher believed, "A company is stronger if it is bound by love rather than by fear." Bob Chapman has discovered that love never fails to help him turn dying companies around. He counsels that you treat every stakeholder as you would your own "precious child." Bob doesn't buy companies to turn them over for a quick profit; he invests in them and "adopts" them forever. Love perseveres and a loving purpose is for the long term.

Inspiring

When people are inspired by a shared purpose, they are aligned in the same direction. In the absence of a shared purpose, stakeholders are literally at cross purposes; shareholders want as much money as possible, employees want to work as little and get paid as much as possible, customers want the lowest price every time, suppliers want to maximize their margins by cutting corners, and society wants to tax the business as much as possible. Everyone becomes a taker from the system.

The inspiring, aligning power of a great purpose preempts many of the conflicts that commonly arise between stakeholders, and enables the discovery of creative, win-win resolutions when conflicts do arise.

A healing purpose inspires all the stakeholders of an enterprise to be *givers* to the system, rising above their self-imposed limitations and striving for the seemingly impossible. Stakeholders retain their distinctive roles and identities, but also voluntarily become part of a synchronized, harmonious whole. This electrifies and galvanizes the organization, giving it a sense of urgency and focus.

Natural

A healing purpose reflects a mindset of living in harmony with nature rather than seeking to conquer or dominate it. Mycologist Paul Stamets notes, "We are all on this planet together, in this time and space. We are all going to die. Everybody we know is going to die. We are going to enter into the fabric of nature from which we sprung. I think that fabric of nature is based on the extension of goodness."[13] Extend goodness by integrating ecological awareness and stewardship into your purpose.

Grounded

A healing purpose is grounded with strategic thinking, careful financial planning, and attention to detail. The twenty-sixth US president, Theodore Roosevelt, advised, "Keep your eyes on the stars, but remember to keep your feet on the ground."[14] Bob Chapman calls it "grounded optimism." He adds, "Leadership is the stewardship of the lives entrusted to us." Are you prepared to be a steward of lives? Healing leaders demonstrate the courage to be idealistic and caring in the face of many pressures to be cynical and selfish. Search your soul to discover your own highest purpose and then find a way to ground it in the way you conduct business.

Once you've defined your purpose for yourself and for your organization, make your commitment to healing and to your core values explicit and tangible to all your stakeholders.

Healing Organizations discover that putting people first gives them a strong advantage. As Chris Hillmann explains, his business is usually very price sensitive, so he was shocked to discover that "the effect of communicating our passion for caring brings something out in clients that makes them want to do business with us and they're willing to pay us more." Jaipur Rugs discovered that customers were willing to pay much more to buy rugs crafted and signed by artisans rather than save money buying from enterprises that exploited indentured slaves. Eileen Fisher finds that investing in sustainability and fair trade inspires the passionate brand loyalty that helps generate consistent profit.

ON BECOMING
A HEALING LEADER

The translation of noble healing ideals into reality requires leadership. You cannot have a Healing Organization without healing leaders. And healing leadership demands a continuing commitment to healing oneself. In healing yourself, you can extend that healing to others.

The amazing individuals we have been privileged to learn from and feature in this book have all evolved through their own life experiences to arrive at a clear sense of themselves and what they are seeking to manifest in the world through their enterprises. In many cases, these leaders have found meaning in their own suffering, and have sought to use the canvas of their businesses to create something that reflects the good, the true, and the beautiful as they have come to understand it.

One lesson they all learned was to become aware of the attitudes, assumptions, and habits that cause unnecessary suffering and change those first.

Healing leaders recognize that dying with the most toys doesn't make you a hero, and that the winner of the rat race is still a rat. A few moments of reflection are all it takes for anyone to realize that, on our deathbeds and afterward, our legacy will be measured by the inspiration, kindness, healing, and love we've shared—and, in business, the contribution we've made to colleagues, customers, suppliers, and our communities.

Once conscience is awakened, the biggest step for many in the journey to becoming a healing leader is understanding that *it is possible* to alleviate suffering and elevate joy through business; and one of the best ways to nurture that understanding is to study the stories of individuals who embody healing leadership. In this book we've shared the stories of some of the exceptional healing leaders that we know, and there are many more. Some are legendary, like Jim Sinegal, founder and former CEO of Costco, who paid his employees nearly double the wages of his direct competitors, and provided excellent benefits. He refused to charge more than a 14 percent markup on anything Costco sold, which meant that, for some generic drugs, Costco's price could be 90 percent below that of its competitors. When pressed by Wall Street analysts to increase its margins on such products, he answered, "I might as well take heroin."[1] Likewise, Howard Schultz of Starbucks resisted all calls, from Wall Street and even from within his own board of directors, to dilute the company's generous health insurance coverage, which he expanded to include part-time employees. Howard had seen firsthand, as a child, the devastating impact that not having health insurance could have on a family. John Mackey of Whole Foods Market set out to create a company that would improve the health of customers, the food system, and the planet. He extended his healing mindset into the realm of animal welfare, by creating the Global Animal Partnership, which has set standards for the humane treatment of animals in the food system.

Sinegal, Schultz, and Mackey are well known, and there are many others who are quietly changing fundamental assumptions about the nature of business. Every industry evolves its own norms, values, and standard practices, which then create a reputation for the industry among potential customers and employees. Think of used-car retailing, timeshare sales, or aluminum siding in the old days. The bad reputation of the industry becomes a stigma that gets attached to every participant in the industry, even new ones. But the worse the industry reputation, the greater the opportunity for a company with a different, healing consciousness to stand out and attract customers, employees, and investors who are looking for something better. In the notoriously hierarchical and often brutal construction industry, Michael Hammes and Barry Dikeman of Ram Construction in Bellingham, Washington, have placed safety and kindness at the center of everything they do and the result is a profitable company that is loved by employees, customers, and the community.

Power Home Remodeling in Chester, Pennsylvania, is committed to changing the way their industry is viewed by all stakeholders, beginning with their mostly millennial employees. Even among industries that do not have a good reputation, the home remodeling industry is exceptionally distrusted. It has a reputation of people who take advantage of homeowners. Common practices are starting jobs, taking money, and never finishing them. Or taking money, starting the job, and then demanding more money to complete it. Customers are misled and treated with disdain, suppliers are not paid as promised, and bills are riddled with unexpected charges. PHR co-CEO Asher Raphael told us, "Our number one goal is to create positive change in the lives of our customers, our employees, and the planet." Power strives to be a "dream realization" company, for its customers as well as for its employees. It has been cited as one of Glassdoor's Best Places to Work. The company's co-CEOs, Corey Schiller and Asher Raphael, have been named to the Glassdoor Top CEO list for the fourth consecutive year. Power has built a culture in which people start believing in themselves, love their work, make good money, and become far more successful in every dimension than they imagined possible.

The company offers growth and development opportunities to all of its employees, not just managers. It has created a culture of inclusion and trust, where people feel psychologically safe and free to be who they are. For example, one employee stood up at the company's first Diversity and Inclusion Summit and with tears in his eyes said, "This the first time I have ever been open about being gay and the first time I have been truly happy in my life. I never thought I could be this happy as a human being before working here."

Healing leaders are transforming business in many realms that are not renowned for their humanistic approach. The Breakers, long known as a haven for the ultra-wealthy families of the gilded age, was patronized by the Rockefellers, Vanderbilts, Astors, Andrew Carnegie, and J. P. Morgan, alongside vacationing European nobility and a parade of US presidents. The hotel was originally built by Henry Flagler (1830–1913). When John D. Rockefeller was asked if creating the massive monopoly that became Standard Oil was his idea, he replied, "No, sir. I wish I had the brains to think of it. It was Henry M. Flagler." Flagler was a genius who helped devise a strategy, based on aggressive acquisitions and the control of rail transport, to dominate the global energy trade. After amassing one of the world's greatest fortunes through his collaboration with Rockefeller, Flagler turned his attention to developing the state of Florida. Like the other robber barons, Flagler devoted much of his energy in his later years to leaving a more positive legacy. He helped to endow the business school at the University of North Carolina—it still bears his name—and he believed that making Florida a viable, functioning state would help strengthen trade with Central America and support mutual prosperity between the US and its southern neighbors. Part of his strategy for transforming Florida was to build fabulous luxury hotels that would attract wealthy guests who would then invest in the local infrastructure. In addition to building his own extraordinary estate, he opened The Royal Poinciana Hotel and then the Palm Beach Inn, which soon changed its name to The Breakers, a reference to the way the waves of the Atlantic Ocean meet the shoreline at this magnificent location.

Over the years the hotel has been rebuilt after fires and hurricanes, but its greatest transformation began in 2003 when Flagler's heirs made a commitment to transform its culture. Prior to 2003, the annual turnover rate for employees was 100 percent. In other words, people on average stayed for only a year! Now they are down to a 17 percent turnover rate, which is one of the best in the hospitality industry. As CEO Paul Leone puts it, "We strive to be an organization of conscience and action, a culture of care and well-being. Our drive to succeed across our diverse businesses is ultimately fueled by our commitment to serve and nurture our team, to build an environment where staff can grow and feel inspired to live a fulfilled life, and to help our community and environment flourish as well. This is deeply embedded in our value system, and so we constantly focus on the quality of life of our employees, community engagement, and service to those in need, and respect for environmental preservation and sustainability."

Human Resources Vice President Denise Bober adds, "We place team member well-being at the core of our strategy . . . this is what drives the satisfaction of customers and clients." Nearly twenty years after embarking on its healing journey, The Breakers has seen dramatic, positive changes in every aspect of its culture and operations. It is one of the most profitable hotels in the US and has been able to increase its average daily rate nearly three times more than its competitor set. Guests at the resort are some of the most satisfied and delighted in the world, as evidenced by its satisfaction scores (90 percent) and repeat business (94 percent).

And there are many, many more stories like these. The first key to becoming a healing leader is to immerse yourself in these stories and share them with others. When we told the story of Appletree Answers' Make-a-Wish program for employees to a friend who is CEO of a global manufacturing company, he became so inspired that he immediately began the implementation of a similar initiative in all his factories.

There are many wonderful guides to leadership that you can read to generate more inspiration and to develop the skills that all healing leaders need, including empathic listening, mindfulness, creativity, conflict management, and negotiation. We needn't reiterate them all here. Rather we will just share guidance based on ten of the elements that are distinctive in the healing leaders we spoke to for this book.

If you want to be a healing leader:

EMBRACE INNOCENCE AND HUMILITY

Being a healing leader doesn't require you to have all the answers but it does demand that you are willing to ask lots of questions. Innocence is powerful, because it lets you ask seemingly naive, childlike questions, which can serve you extremely well. When Mark Hoplamazian took over as the CEO of Hyatt Hotels, he didn't know much about the business. Because he knew that he didn't know and was curious to learn, he asked simple questions that yielded new perspectives. These encouraged deep engagement from his team—leading, ultimately, to many positive changes and healing innovations. Innocence is not only about not knowing or assuming; it is also about never knowingly and without justification hurting another. N. K. Chaudhary sees himself and his 40,000 weavers as such innocents; he believes that "the cunning of others is our greatest opportunity." This quality is what is referred to in the Beatitudes as "meek" or "pure of heart" and it is a key source of healing power.

ALCHEMIZE YOUR OWN SUFFERING

All of us have had traumas in our lives, some more severe than others. For most people, the effects of those traumas linger for a lifetime, distorting their perceptions, shaping their behavior, and draining their lives of fulfillment and happiness. Healing leaders recognize that you cannot change the past, nor can you forget it; but you can learn from it to heal yourself and help others. They seek to find meaning in the suffering they may have experienced early in life and break the cycle of victimhood. They are able to give what they never got. Eileen Fisher, for example, felt uncomfortable in her own skin, so she created a business that makes clothes that help women feel comfortable with themselves, and an organizational culture that supports team members in feeling whole and connected. Danny Meyer struggled with conflict and dysfunction in his family, so he created a working environment that models many of the best elements of a healthy family system.

Healing leaders organize what they do around the principles of caring and kindness. Like Daniel Lubetzky, they know the difference between superficial niceness and genuine kindness. They recognize that they and their teams are on a never-ending journey to embody their highest ideals. "I don't know that we fully live those principles," comments Danny Meyer, "but having the ideals gets you closer."

BE TRUE TO YOURSELF

It is fine to seek role models, but sooner or later we realize that we must discover our own unique path. Others can inspire us and help us realize what's possible, but ultimately we need to find our own original essence and live from our own deepest truths. As Tami Simon told us, "There is no one else that's you anyway. No one else has your exact DNA, your parents, your exact combination of promises you made when you came here to deliver to the world."

Being true to yourself becomes more challenging as companies grow, and especially if they become public. There are numerous forces that demand conformity and offer a nice premium for your soul. As you become more successful, the temptation to sell your soul gets greater. Florence Butt defied convention to go to college, graduate as the only woman in her class, and then start a business. She managed to endow her son and grandson with the spiritual wherewithal to maintain and extend the "Spirit of Giving" that makes H-E-B an exemplary Healing Organization.

Be vigilant and focused on your truth and the truth of your enterprise's higher purpose. As Tami Simon explains, "I have a thin line I can walk for my soul to be intact. If I veer off of it, things aren't good for me. And when I do stay on it, my life unfolds beautifully." Tami adds, "Everybody kind of knows what they need for their souls to be healthy."

MODEL THE VALUES AND BEHAVIORS YOU WISH THE ORGANIZATION TO HAVE

Although we are all unique and must discover our originality, you can be sure that if you are in a leadership position others will be modeling themselves after you, for better or for worse. So be impeccable about manifesting whatever you want to inculcate in the organization. As David Gardner told us, "We love our kids into loving and smile our babies into smiling. Our parents trusted us into trusting. People who lie start to believe that others are lying to them. People who do bad things in this world live in fear that other people may be doing similar things to them."

Over the years, Eileen Fisher has interrupted many meetings to attend to the needs of her autistic child. By publicly putting her family first in this way, she made it possible for other people at the company to do the same.

The good news is that healing begets healing. It sets in motion a virtuous cycle. By modeling care, you inspire caring. By helping others, you help yourself. By creating an environment of healing, you heal yourself as well.

As a leader, always ask, "Are we doing enough—for each other and for all our stakeholders?" Leaders adopt a "continuous improvement" approach toward a healing orientation; they never forget that, *there is always a better way.*

THINK CREATIVELY AND LEAD INNOVATION

Creativity is the art of generating new ideas that have subjective value, and innovation is the process of translating those ideas into objective value, i.e., products or services that meet human needs in a sustainably profitable way. Healing leaders are solution-oriented, creative thinkers who inspire and guide others to find creative solutions, thus nurturing a culture that supports innovation. Like Daniel Lubetzky, they understand how to frame challenges as opportunities to move beyond limiting either/or constructs (*either* healthy *or* delicious) and instead apply both/and solutions (healthy *and* delicious).

BECOME AN INSPIRING STORYTELLER

The Healing Organization is a new dream, a fresh story, an evolving myth about business. As the shaman Numi explained to John Perkins, "All you have to do is change the dream. It can be accomplished in a generation. You need only plant a different seed, teach your children to dream new dreams."[2]

Storytelling is the way new seeds are planted for a more beautiful dream to emerge and thrive. Storytelling defines or redefines an organization's culture and reinforces its values and purpose. Everybody at Barry-Wehmiller knows "The Wedding Story." They also know how Randy Fleming became Randall Fleming. The stories about what H-E-B did in the aftermath of Hurricane Harvey are legendary. Menlo Innovation founder Richard Sheridan's title is *Chief Storyteller*.

REMEMBER THAT MEANS AND ENDS ARE INSEPARABLE

All of our healing leaders realized that the *way* you do what you do is as important as what you are doing. They want employees and their families to thrive because that is the right thing for them, not so that they can get more out of them. They recognize there is no real end; each "end" is just the means to something else. As the Bhagavad Gita explains:

To action alone hast thou a right and never at all to its fruits; let not the fruits of action be thy motive; neither let there be in thee any attachment to inaction.[3]

In other words, "The wise are not bound by desire for rewards." As Tami Simon expressed it: "The means, the path itself, should always be life-giving."

HARMONIZE THE FOUR ARCHETYPAL ENERGIES

The leaders we spoke with are men and women, young and old, straight and gay, but all seek to harmonize the four archetypal energies in themselves and their organizations. They all manifest the elder energy by organizing their enterprises around a higher purpose based on universal wisdom. They balance the masculine and feminine energy with a powerful integration of analytical, assertive, action-orientation with patience, caring, and imagination. And they all celebrate the healthy child energy by encouraging a playful and joyful way of being at work.

THINK AHEAD FOR SEVEN GENERATIONS—AT LEAST!

The chiefs of the Iroquois Confederacy understood that our thoughts and actions today set the stage for the world our children and our children's children and all our ultimate descendants will inherit. A healing leader considers the multigenerational impact of the decisions made now on the future of humanity.

In the words of legendary investor Warren Buffett, "Someone's sitting in the shade today because someone planted a tree a long time ago."[4] If you are an American then, (unless you are from an original tribe) you must be grateful that your parents, grandparents, great-grandparents, or more distant ancestors had the courage and foresight to emigrate so that you could be free to enjoy the benefits of democracy and capitalism.

Thirty years ago in Vermont, a company was formed "to inspire a consumer revolution that nurtures the health of the next seven generations." The Seventh Generation company was formed because a group of dedicated entrepreneurs "looked at the world around them and saw it heading in the wrong direction." As CEO Joey Bergstein explains, "There was too much dirty air and not enough clean water, too many toxins and no real alternatives, too much waste and not enough wisdom."

Recently acquired by Unilever, the company has taken on an even greater healing mission: "To demonstrate that doing no harm is just the start. The real aim can and must be for all of us to live and work in ways that actually leave the world better off than it was when we arrived."

ALWAYS OPERATE FROM LOVE

The healing leaders we interviewed come from many different spiritual traditions—Christians, Jews, Muslims, Buddhists, Hindus, and Atheists/Humanists—but they all believe that love, rather than fear, must be at the center of our lives and our businesses. As Nand Kishore Chaudhry states, "Leaders driven by love will bring sustainability and healing to the business, as well as for themselves."

They share a sense that our planet is one system and that our fates are intertwined. They understand, as Leonardo da Vinci observed more than five hundred years ago, that: "Everything is connected to everything else."[5]

The consciences of the healing leaders we interviewed were awakened either through experiencing or witnessing suffering, or from an overflow of abundance that leads to an outpouring of gratitude. In all cases, their work has become an expression of Love.

Work as an Expression of Love

(adapted from *The Prophet* by Kahlil Gibran)[6]

Loving life through labor is to be intimate with life's innermost secret.

All work is empty save when there is love; for work is love made visible.

And when you work with love you bind yourself to yourself, and to one another, and to God.

What is it to work with love?

It is to weave the cloth with threads drawn from your heart, even as if your beloved were to wear that cloth. (like Eileen Fisher and Jaipur Rugs)

It is to build a house with affection, even as if your beloved were to dwell in that house. (like Hillmann Consulting and Hyatt)

It is to sow seeds with tenderness and reap the harvest with joy, even as if your beloved were to eat the fruit. (like KIND Healthy Snacks and USHG)

It is to charge all things you fashion with a breath of your own spirit. (like *Conscious Company* magazine, H-E-B, DTE,—indeed all Healing Organizations)

Our capacity to give and receive love is what defines us as human beings, as divine creatures endowed with intelligence, imagination, and free will. If we wish, it can pervade our lives with sweetness, beauty, and joy. But too often we keep love confined and limited. We dole it out carefully and stingily, as though we were parting with a severely limited resource. As a result, our lives become mechanistic, arid, and devoid of joy.

We beseech you: Always operate from love!

This is what Ramon Mendiola did when he vowed to his employees that not a single one of them would have to contend with the ravages of poverty. It is what Gerry Anderson did when he set out to extend the healing he had wrought within the company to the devastated Detroit region. This is what Eileen Fisher does when she supports her people in their personal growth and inner healing.

Love keeps hope alive, even for people whose circumstances seem hopeless. There's a nineteenth-century African American spiritual hymn that goes:

> There is a balm in Gilead
> To make the wounded whole;
> There is a balm in Gilead
> To heal the sin-sick soul.

The balm is love. And you don't have to travel to Gilead to find it.

Now, as you embrace the journey of realizing that business *can be* and *must be* and is fundamentally *meant to be* about healing, make this your mantra: *Operate from love.*

EPILOGUE

IMAGINE A NEW WORLD

The premise of a Healing Organization is simple:

> When we understand and meet people's *real* needs, we help to heal them, while healing ourselves and generating abundance. When, instead, we uncover and prey on their cravings, desires, fears, and addictions, we hurt them, and ultimately we hurt ourselves, our children, and our planet.

Inspired by the US Declaration of Independence, the United Nations adopted a Universal Declaration of Human Rights in 1948. Article 1 states:

> All human beings are born free and equal in dignity and rights. They are endowed with reason and conscience and should act toward one another in a spirit of brotherhood.

Imagine what would happen to our world if we transformed our way of thinking about business so that it was based on a spirit of brotherhood that elevated reason and conscience and promoted freedom and equality.

Imagine a world in which business makes human flourishing its first priority.

Imagine what that would mean for the mental, physical, emotional, and spiritual well-being of people at work—and for their children, families, and communities.

Imagine the consequences for the quality of the air we breathe, the water we drink, and the land and seascapes that support our lives.

As more businesses embrace the path of healing:

We believe that depression, anxiety, addiction, and suicides will decline and become rare.

We believe that the walls separating labor and management will crumble as all realize that creative caring collaboration yields more fulfillment and well-being for all parties.

We believe that our political divides will begin to ease as the values that unite us—liberty, prosperity, dignity, fairness, opportunity, and love—come to life in our workplaces.

The time for Healing is *now*.

Now is the time to redefine success so that it includes and promotes the values and traits that we cherish. For this to happen we must all share stories of healing business and celebrate real heroes rather than glorifying predators.

Former US president Franklin D. Roosevelt observed:

Human kindness has never weakened the stamina or softened the fiber of a free people. A nation does not have to be cruel to be tough.[1]

Nor does a business.

Former US Attorney General Robert F. Kennedy offered a perspective that lives on fifty years after his assassination:

Few will have the greatness to bend history itself, but each of us can work to change a small portion of events. It is from numberless diverse acts of courage and belief that human history is shaped. Each time a man stands up for an ideal, or acts to improve the lot of others, or strikes out against injustice, he sends forth a tiny ripple of hope, and crossing each other from a million different centers of energy and daring those ripples build a current which can sweep down the mightiest walls of oppression and resistance.[2]

THE HEALING ORGANIZATION OATH.

We see this book as part of a movement to change the world of business and make it about love and healing instead of fear and survival. If you'd like to be part of this movement, begin by taking The Healing Organization Oath.

Place your left hand on your heart and raise your right hand and proclaim:

Primum non nocere (First, do no harm).

I will operate my business in a way that causes no harm to others or to the earth.

Malus eradicare (Root out evil).

I will never enable or collude with abuse or exploitation. I will be an everyday hero who stands up for fairness, truth, beauty, integrity, and basic goodness.

Amor vincit omnia (Love conquers all).

I will operate from love. I will measure success by the fulfillment, abundance, and joy I generate for others.

ACKNOWLEDGMENTS

RAJ

I am grateful to my sage and loving friends Nilima Bhat, Louisa and Ilan Bohm, and Lynne Twist, who helped guide my personal healing journey and also counseled me to take time to go inward and learn more about healing and how to heal myself in preparation for writing this book. We delayed the book by several months so that I could experience a silent retreat in upstate New York, a spiritual sojourn in Ladakh in the high Himalayas, and a trip deep into the Amazon rain forest with the Pachamama Alliance that included several shamanic healing experiences. Every one of these experiences brought forth profound insights and revelations into the meaning of life and the true nature of business, and how both are ultimately bound by love.

I'm deeply grateful to my mother, Usha, for her unconditionally loving, healing presence in my life from the beginning. My sister, Manjula, and her husband, Sangram, gently guided me toward spiritual exploration. I'm grateful to my dynamic and fearless brother, Sanjay, who is taking on the monumental challenge of healing our ancestral lands in India. My cousin Gajendra has shown me from the day I was born what innocence and unconditional love look like.

My mentor Jag Sheth, and his wife, Madhu, have been like parents to me and gave me courage to follow my own path. Jag taught me more about business and life than anyone else.

My friends John Mackey, Doug Rauch, and Kip Tindell inspire me as models of conscious leadership, as does Bob Chapman, who helped give birth to the idea that businesses should grow like ministries instead of empires. Haley Rushing and Roy Spence have enriched my thinking about purpose and healing. Betsy Sobiech and Alexander Mendeluk contributed important insights on our journey together to the Himalayas.

I received priceless wisdom during my Pachamama Alliance trip from John Perkins and my fellow travelers to the rain forest, especially Sara Vetter and David Applefield, and our peerless guide, Daniel Koupermann.

I thank Sandra Waddock for including me in a book titled *Intellectual Shamans*, which prompted me to think of myself as a healer in the world of business. My coach Suzanne Vaughn helped me clarify my purpose and make sense of my life. Thanks also to Peter Senge for hosting the marvelous silent retreat that awakened me to many insights that informed this work, and to the brilliant Neha Sangwan for sharing her healing wisdom and practices.

Finally, I would like to thank my coauthor, the extraordinary polymath Michael J. Gelb, and his wonderful wife, the radiant mezzo-soprano Deborah Domanski. Writing this book with Michael was sheer delight; he is a supremely gifted thinker, writer, speaker, and teacher in numerous realms, a true Renaissance man. I am in awe of his manifold gifts and am grateful to be able to call him my friend. Deborah has been along for every step of this journey, warmly welcoming me to their love-filled home on our numerous writing retreats.

Michael and Deborah, I love you and I thank you.

MICHAEL

Rumi referred to gratitude as "wine for the soul." My cup runneth over with thanks to all who supported and inspired me in this journey including Stephanie Cesario, Deborah Domanski, Andrew Dornenburg, Joan and Sandy Gelb, Chris Hillmann, Matt Kamin, Jon Miller, Ashley Munday, Karen Page, and David Saltman.

G. K. Chesterton referred to gratitude as "happiness doubled by wonder." I couldn't be happier with my wonderful coauthor, Raj Sisodia. Raj is much more than just a gifted teacher and thought leader; he is a planetary *soul-force* hero, a champion of truth, beauty, and goodness. I'm honored to call him my friend and my brother. Raj, we love you and we thank you.

RAJ AND MICHAEL

We thank Nukul Jain and Siddharta Ramachandran for their research assistance. We are grateful to Ellen Kadin for supporting the vision, and to our brilliant team from HCL: Sicily Axton, Tim Burgard, Hiram Centero, and Sara Kendrick.

We are deeply grateful to the exemplary leaders and thinkers who generously shared their wisdom and stories with us:

Shawn Achor, Adyashanti, Gerry Anderson, Richard Barrett, Howard Behar, Denise Bober, Craig Boyan, Mike Brady, John Hope Bryant, Bob Chapman, Nand Kishore Chaudhry, David Cooperrider, Anthony Cullwick, Barry Dikeman, Meghan French Dunbar, Jane Dutton, Eileen Fisher, Lawrence Ford, David Gardner, Tom Gardner, Bernie Glassman, Jonathan Haidt, Michael Hammes, Doug Hammond, Chris Hillmann, Mark Hoplamazian, Joseph Jaworski, Kristine Jordan, Herb Kelleher, Fred Kofman, Frederic Laloux, Bill Linton,

ACKNOWLEDGMENTS

Daniel Lubetzky, Nipun Mehta, Danny Meyer, Asher Raphael, John Ratliff, Chris Reinking, Ramon Mendiola Sanchez, Tony Schwartz, Dov Seidman, Safwan Shah, Tal-Ben Shahar, Casey Sheahan, Richard Sheridan, Tami Simon, Roy Spence, Lynne Twist, Marianne Williamson, Steev Wilson, Monica Worline, and Nigel Zelcer.

ENDNOTES

PROLOGUE

1. Jagdish N. Sheth, Rajendra S. Sisodia, and Adina Barbulescu, "The Image of Marketing," in Jagdish N. Sheth and Rajendra S. Sisodia (eds.), *Does Marketing Need Reform? Fresh Perspectives on the Future* (New York: Routledge, 2006).
2. "The Harder Hard Sell: The Future of Advertising," *The Economist*, June 24, 2004.
3. Elisabeth Leamy, "How to Stop Junk Mail and Save Trees—and Your Sanity," *Washington Post*, February 14, 2018.
4. For a broader list, see https://www.consciouscapitalism.org/boardofdirectors.

INTRODUCTION

1. https://en.wikiquote.org/wiki/Homer.
2. Charles Dickens, *A Tale of Two Cities* (1859).
3. "Homer vs. the Eighteenth Amendment," *The Simpsons*, Season 8, Episode 18. Originally aired March 16, 1997.
4. https://similarworlds.com/32-Personal-Thoughts-Feelings/1738203-Im-not-going-to-stop-torturing-myself-until-I.
5. Dr. Eger spoke at Sages & Scientists 2014, https://www.choprafoundation.org/speakers/edith-eva-eger-phd/.
6. Emi Suzuki and Haruna Kashiwase, "New Child Mortality Estimates show that 15,000 Children Died Every Day in 2016," *The World Bank Data Blog*, October 19, 2017, https://blogs.worldbank.org/opendatanew-child-mortality-estimates-show-15000-children-died-every-day-2016.
7. Alexa Lardieri, "World Bank: Half the World Lives on Less Than $5.50 a Day," *U.S. News & World Report*, October 17, 2018, https://www.usnews.com/news/economy/articles/2018-10-17/world-bank-half-the-world-lives-on-less-than-550-a-day.

8.   https://www.paloaltoonline.com/news/2017/03/03/cdc-releases-final-youth
     -suicide-report; https://www.nationalgeographic.com/adventure/adventure-blog
     /2016/05/16/why-are-ski-towns-suicides-happening-at-such-an-alarming-rate/.

9.   Michael J. Gelb, *How to Think Like Leonardo da Vinci: Seven Steps to Genius
     Every Day* (New York: Bantam Dell, 2004), 226.

10. Walter Sullivan, "The Einstein Papers. A Man of Many Parts," *New York Times*,
     March 29, 1972.

11. David Loye, "To Darwin: A Birthday Manifesto," *AntiMatters* 3:1 (2009)
     https://antimatters2.files.wordpress.com/2018/04/3-1-loye.pdf.

12. Muhammad Yunus, "Redesigning Economics to Redesign the World," *The
     Daily Star*, November 18, 2014, https://www.thedailystar.net/redesigning
     -economics-to-redesign-the-world-50798.

13. Louis Menand, "Karl Marx, Yesterday and Today," *The New Yorker*, October
     10, 2016, https://www.newyorker.com/magazine/2016/10/10/karl-marx-yesterday-
     and-today.

14. "Guolaosi," Schott's Vocab: A Miscellany of Modern Words and Phrases, *New
     York Times*, August 18, 2010, https://schott.blogs.nytimes.com/2010/08/18
     /guolaosi/.

15. Anahad O'Connor, "The Claim: Heart Attacks Are More Common on Mon-
     days," *New York Times*, March 14, 2006, https://www.nytimes.com/2006
     /03/14/health/14real.html.

16. Jeffrey Pfeffer, *Dying for a Paycheck: How Modern Management Harms Em-
     ployee Health and Company Performance—and What We Can Do About It*
     (Harper Business, 2018).

17. Horace Mann, *Twelve Sermons Delivered at Antioch College* (Boston: Ticknor
     and Fields, 1861), p. 182.

18. Henry Cabot Lodge, *The Life of George Washington*, vol. 1 (Boston: Houghton
     Mifflin, 1920), p. 52.

19. Charmaine Li, *Confronting History: James Baldwin*, https://kinfolk.com
     /confronting-history-james-baldwin/.

20. Both authors serve on the advisory board of LifeGuides, and Raj is an investor
     in the company.

21. www.LifeGuides.com; personal communication from Mark Donohue.

22. Charles Eisenstein, *The More Beautiful World Our Hearts Know Is Possible*
     (Berkeley: North Atlantic Books, 2013).

23. Robert Wright, *Nonzero: The Logic of Human Destiny* (Pantheon, 1999).

24. Margaret Moodlan, "Lessons of Compassion from the Dalai Lama," The Blog,
     Huffpost, July 24, 2015, https://www.huffingtonpost.com/margaret-moodian
     /lessons-of-compassion-fro_b_7868940.html.

25. Dickens, *A Tale of Two Cities*, book 3, chapter XV.

26. Dickens, *A Tale of Two Cities*, book 3, chapter XV.

RE-DREAMING THE AMERICAN DREAM

1. "America Is a Nation with the Soul of a Church," *The Apostolate of Common Sense* (blog), April 30, 2012, https://www.chesterton.org/america/.
2. "Iroquois Confederacy," Native American Tribes site, Siteseen Limited, November 20, 2012 (updated January 16, 2018), https://www.warpaths2peacepipes .com/native-american-indians/iroquois-confederacy.htm.
3. Molly Larkin, "The History of the U.S. Constitution We Weren't Taught in School," blog post, July 2, 2012, https://www.mollylarkin.com/the-history -u-s-constitution-we-werent-taught-school/.
4. "Iroquois Constitution: A Forerunner to Colonists' Democratic Principles," *New York Times*, June 18, 1987, https://www.nytimes.com/1987/06/28/us /iroquois-constitution-a-forerunner-to-colonists-democratic-principles.html.
5. Louis Jacobson, "Viral Meme Says Constitution 'Owes Its Notion of Democracy to the Iroquois,'" PolitiFact, December 2, 2014, https://www.politifact .com/truth-o-meter/statements/2014/dec/02/facebook-posts/viral-meme-says -constitution-owes-its-notion-democ/.
6. This encounter was described in Walter Isaacson's biography of Benjamin Franklin, *Benjamin Franklin: An American Life* (New York: Simon & Schuster, 2004).
7. Rob Wile, "The True Story Of The Time JP Morgan Saved America From Default By Using An Obscure Coin Loophole," *Business Insider*, January 13, 2013, https://www.businessinsider.com/morgan-1895-crisis-and-1862-gold-loophole -2013-1; "Panic of 1907: J.P. Morgan Saves the Day," https://www.u-s-history .com/pages/h952.html.
8. Peter Krass, *Carnegie* (Hoboken: Wiley, 2001).
9. From the documentary *The Men Who Built America*, season 1, episode 7, "Taking the White House," counter 00:03:48.
10. Megan Day, "Andrew Carnegie Once Hired a Militia and Converted Factories into Makeshift Forts to Battle Striking Workers," Timeline, February 8, 2018, https://timeline.com/dale-carnegie-militia-battle-striking-workers-c0fdc8a 75527.
11. "The Strike at Homestead Mill," *Andrew Carnegie: The Richest Man in the World*, American Experience site, PBS/WGBH, https://www.pbs.org/wgbh /americanexperience/features/carnegie-strike-homestead-mill/.
12. Robert Sobel, "Coolidge and American Business," Calvin Coolidge Presidential Foundation, 1988, https://www.coolidgefoundation.org/resources/essays-papers -addresses-35/.
13. https://www.azquotes.com/author/13714-Alfred_P_Sloan.
14. "'Warrior' Spirit," Southwest blog, March 18, 2008, https://www.southwestair community.com/t5/Southwest-Stories/quot-Warrior-quot-Spirit/ba-p/43921.
15. John Perkins, *The World Is As You Dream It: Teachings from the Amazon and Andes* (Rochester, VT: Destiny Books, 1994).

16. C. J. Green, "Martin Luther King, Jr: A Tough Mind and a Tender Heart," *Mockingbird*, January 16, 2017, https://www.mbird.com/2017/01/martin-luther-king-jr-a-tough-mind-and-a-tender-heart/.

17. Holly Hedegard, Sally C. Curtin, and Margaret Warner, "Suicide Mortality in the United States, 1999–2017," National Center for Health Statistics, NCHS Data Brief No. 330, November 2018, https://www.cdc.gov/nchs/products/data-briefs/db330.htm; "Suicide Is Declining Almost Everywhere," *The Economist*, November 24, 2018, https://www.economist.com/international/2018/11/24/suicide-is-declining-almost-everywhere.

18. George Packer, *The Unwinding: An Inner History of the New America* (New York: Farrar, Straus and Giroux, 2014).

19. For example: Per capita incomes have risen approximately 1,500 percent since the year 1800. Literacy has gone from 13 percent to 87 percent. Life expectancy has risen from approximately 32 percent to approximately 72 percent. The percentage of people living in extreme poverty has declined from 90 percent to below 9 percent. See John Mackey and Raj Sisodia, *Conscious Capitalism: Liberating the Heroic Spirit of Business* (Boston: Harvard Business Review Press, 2014).

20. See www.firmsofendearment.com for details.

EVOLVING FROM EMPIRE TO MINISTRY, FROM CONQUERING TO CARING

1. Steven Pinker, *The Better Angels of Our Nature* (New York: Viking, 2011).

2. Colin Dodds, "Kevin O'Leary: Most Influential Quotes," Investopedia, n.d., https://www.investopedia.com/university/kevin-oleary-biography/kevin-oleary-most-influential-quotes.asp.

3. Colin Dodds, "Kevin O'Leary."

4. Jim Collins, *Good to Great: Why Some Companies Make the Leap . . . and Others Don't* (New York: Harper Business, 2001).

5. "Heavy Smokers Cut Their Lifespan by 13 Years on Average," Central Bureau voor de Statistiek (NL), September 15, 2017, https://www.cbs.nl/en-gb/news/2017/37/heavy-smokers-cut-their-lifespan-by-13-years-on-average; "Tobacco," World Health Organization, March 9, 2018, https://www.who.int/news-room/fact-sheets/detail/tobacco; "Diseases Linked to Smoking Cost the World $422 Billion in Health-related Expenses," American Cancer Society, January 31, 2017, https://www.cancer.org/latest-news/diseases-linked-to-smoking-cost-the-world-422-billion-in-health-related-expenses.html.

6. https://english.stackexchange.com/questions/444923/is-there-a-common-saying-in-english-that-means-its-just-business-i-dont-feel/444944.

7. https://www.unilever.ca/about/.

8. Kim Bhasin, "Unilever CEO: The Very Essence of Capitalism Is Under Threat," *Business Insider*, June 22, 2012, https://www.businessinsider.com/unilever-ceo-paul-polman-the-very-essence-of-capitalism-is-under-threat-2012-6.

9. Daniel Roberts, "Here's What Happens When 3G Capital Buys Your Company," *Fortune*, March 25, 2015, http://fortune.com/2015/03/25/3g-capital-heinz -kraft-buffett/.

10. Roberts, "Here's What Happens."

11. Jo Confino, "Unilever's Pail Polman: Challenging the Corporate Status Quo," *The Guardian*, April 24, 2012, https://www.theguardian.com/sustainable -business/paul-polman-unilever-sustainable-living-plan.

12. Leila Abboud and Camilla Hodgson, "Unilever Chief Paul Polman to Step Down," *Financial Times*, November 29, 2018, https://www.ft.com/content/4fd75572 -f3a6-11e8-9623-d7f9881e729f.

13. Jon Ronson, "Your Boss Actually Is a Psycho," *GQ*, December 19, 2015, https ://www.gq.com/story/your-boss-is-a-psycho-jon-ronson.

14. John A. Byrne, *Chainsaw: The Notorious Career of Al Dunlap in the Era of Profit-at-Any-Price* (New York: HarperBusiness, 1999).

15. Byrne, *Chainsaw*.

16. Matt Egan, "5,300 Wells Fargo Employees Fired Over 2 Million Phony Accounts," CNN Business, September 9, 2016, https://money.cnn.com /2016/09/08/investing/wells-fargo-created-phony-accounts-bank-fees /index.html.

17. Sam Harris on the podcast *Armchair Expert*, hosted by Dax Shepard, Episode 58, aired November 22, 2018.

18. Yunus, "Redesigning Economics to Redesign the World."

19. Stanley Milgram, *Obedience to Authority: An Experimental View* (New York: HarperCollins, 1974).

20. https://www.youtube.com/watch?v=wdUu3u9Web4.

21. Philip Zimbardo, "The Psychology of Evil," TED talk, February 2008, https ://www.ted.com/talks/philip_zimbardo_on_the_psychology_of_evil?language =en.

22. Zimbardo, "The Psychology of Evil."

23. Saul McLeod, "The Milgram Shock Experiment," Simply Psychology (updated 2017), https://www.simplypsychology.org/milgram.html.

24. https://condenaststore.com/featured/im-neither-a-good-cop-nor-a-bad-cop-mick -stevens.html.

25. Zimbardo, "The Psychology of Evil."

26. https://www.heroicimagination.org/.

27. Christian Violatti, "Ashoka the Great," Ancient History Encyclopedia, April 11, 2018, https://www.ancient.eu/Ashoka_the_Great/.

28. Vincent Arthur Smith, *Asoka: The Buddhist Emperor of India* (New York: Oxford University Press, 1920).

29. Nayanjot Lahiri, *Ashoka in Ancient India* (Boston: Harvard University Press, 2015), pp. 20–21.

30. Lahiri, *Ashoka in Ancient India*, pp. 20–21.

31. https://en.wikipedia.org/wiki/Herbert_Spencer.

32. Joseph Frazer Wall, *Andrew Carnegie* (Pittsburgh: University of Pittsburgh Press, 1989), p. 386.

THE POWER OF INNOCENCE

1. See, for example, Maria Gonzalez, *Mindful Leadership: The 9 Ways to Self-Awareness, Transforming Yourself, and Inspiring Others* (Mississauga: Jossey-Bass, 2012).

THE ZEN OF BROWNIES

1. https://zenpeacemakers.org/2017/03/sixty-year-journey-2014-dharma-talk-bernie-glassman-eve-marko/.
2. Dan Clark, "How Many U.S. Adults Have a Criminal Record? Depends on How You Define It," Politifact, August 18, 2017, https://www.politifact.com/new-york/statements/2017/aug/18/andrew-cuomo/yes-one-three-us-adults-have-criminal-record/.

THE PARABLE OF THE POTHOLE

1. Anna Bahney, "40% of Americans Can't Cover a $400 Emergency Expense," CNN Money, May 22, 2018, https://money.cnn.com/2018/05/22/pf/emergency-expenses-household-finances/index.html.

MAKING ROOM FOR DREAMS

1. Data provided by PayActiv.
2. Barbara Ehrenreich, "Preying on the Poor: How Government and Corporations Use the Poor as Piggy Banks," *The Nation*, May 17, 2012, https://www.thenation.com/article/preying-poor/. Originally published in TomDispatch.com.
3. Raj serves on the advisory board of PayActiv and owns shares in the company.

THANKS FOR PUTTING POISON ON MY MICROSCOPE

1. Michael has served as an advisor to Hillmann Consulting for the past few years.

ENDNOTES

WHERE ARE THE CUSTOMERS' YACHTS?

1. Fred Schwed, *Where Are the Customers' Yachts?: or A Good Hard Look at Wall Street* (New York: Simon & Schuster, 1940).
2. Laura Blumenfeld, "Voodoo Economics: A Financial Planner Turns Shaman to Manage His Clients' Money and Their Souls," *Washington Post Magazine*, December 7, 2008.

NOT "ONLY" FOR PROFIT

1. Andrew Ross Sorkin and Michael J. de la Merced, "Snickers Owner to Invest in KIND, Third-Biggest Maker of Snack Bars," *New York Times*, November 29, 2017.

HOW *DO* YOU GET TO CARNEGIE HALL?

1. http://www.consultingmag.com/top-25-consultants/.

ENLIGHTENED HOSPITALITY

1. https://www.unionsquarecafe.com/page/restaurant/.
2. Lisa Fickenscher, "How to Close a Restaurant," *Crain's New York Business*, October 31, 2010, https://www.crainsnewyork.com/article/20101031/SMALL BIZ/310319978/how-to-close-a-restaurant.
3. Thomas Oppong, "Pygmalion Effect: How Expectation Shapes Behavior for Better or Worse," Medium, August 1, 2018, https://medium.com/@alltop startups/pygmalion-effect-how-expectation-shape-behaviour-for-better-or -worse-11e7e8fa7f4b.
4. https://www.gramercytavern.com/about/.
5. Brandon Gaille, "23 Greatest Danny Meyer Quotes," *Brandon Gaille Small Business & Marketing Advice*, November 5, 2015, https://brandongaille .com/23-greatest-danny-meyer-quotes/.

FROM DEATH MARCH TO JOY RIDE

1. David Laws, ed., "1971: Microprocessor Integrates CPU Function onto a Single Chip," *The Silicon Engine*, Computer History Museum, https://www.computer history.org/siliconengine/microprocessor-integrates-cpu-function-onto-a -single-chip/.

2. "Edison's Patents," Thomas A. Edison Papers, Rutgers, the State University of New Jersey, last updated October 28, 2016, http://edison.rutgers.edu/patents .htm.

FROM ELEGY TO EXULTATION

1. J. D. Vance, *Hillbilly Elegy: A Memoir of a Family and Culture in Crisis* (New York: Harper, 2016).

THE SPIRIT OF GIVING

1. https://careers.heb.com/our-culture/.
2. https://abc13.com/science/crushing-weight-of-harvey-flood-pushed-houston -down/2413363/; https://www.worldvision.org/disaster-relief-news-stories /hurricane-harvey-facts.
3. Chip Cutter, "The Inside Story of What it Took to Keep a Texas Grocery Chain Running in the Chaos of Hurricane Harvey," https://www.linkedin.com/pulse /inside-story-what-took-keep-texas-grocery-chain-running-chip-cutter/.
4. https://therivardreport.com/charles-butt-takes-the-pledge-its-the-right -thing-to-do/.
5. https://www.expressnews.com/business/local/article/H-E-B-to-give-employees -an-ownership-stake-6605442.php.
6. https://www.bizjournals.com/houston/news/2018/01/24/7-houston-companies -named-on-forbes-new-best.html.

THE MEANS AND THE ENDS ARE THE SAME

1. Joe Herring Jr., "Florence Thornton Butt—A History of HEB Grocery," *Comanche Trace Blog*, August 1, 2011, http://www.comanchetrace.com/florence -thornton-butt-a-history-of-heb-grocery.

BRINGING MORE PURA TO THE CORPORATE VIDA

1. https://www.govisitcostarica.com/travelInfo/nationalParks.asp.
2. "Country Facts," Permanent Mission of Costa Rica to the United Nations, https://www.un.int/costarica/costarica/country-facts.
3. From interview with Ramon Mendiola Sanchez, November 5, 2018.
4. Interview with Ramon Mendiola Sanchez.
5. Jessica I. Montero Soto, "Fifco lidera ranking de mejores lugares para trabajar en Costa Rica," *El Financiero*, May 5, 2017, https://www.elfinancierocr.com

/gerencia/fifco-lidera-ranking-de-mejores-lugares-para-trabajar-en-costa-rica
/56WHT7TJE5DFFHX5IHQG7AF4Q4/story/.

CEO = CHIEF EMPATHY OFFICER

1. https://www.britannica.com/topic/Pritzker-family.
2. https://about.hyatt.com/en/hyatthistory.html.
3. David Gelles, "Mark Hoplamazian of Hyatt Hotels on Airbnb and Why Stupid Questions Are Smart," *New York Times*, October 19, 2018, https://www .nytimes.com/2018/10/19/business/mark-hoplamazian-of-hyatt-hotels-on -airbnb-and-why-stupid-questions-are-smart.html.
4. Gelles, "Mark Hoplamazian of Hyatt Hotels."
5. Damanick Dantes, "Hyatt Hotels CEO: Kindness and Empathy 'Translate Into Good Business,'" *Fortune* (Members Only series), August 1, 2018, http://fortune .com/2018/08/01/hyatt-hotels-ceo-mark-hoplamazian/.
6. Interview with Mark Hoplamazian, December 19, 2017.

THREE PRINCIPLES THAT DEFINE A HEALING ORGANIZATION

1. Linda Rutherford, "Farewell to Southwest's Founder," *Southwest Stories Blog*, January 3, 2019, https://www.southwestaircommunity.com/t5/Southwest -Stories/Farewell-to-Southwest-s-Founder/ba-p/84481.
2. Frank Pierce Jones, "Dewey and Alexander," *Freedom to Change: The Development and Science of the Alexander Technique* (Berkeley: Mornum Time Press, 1997). Originally published as *Body Awareness in Action* (1976), http://www. alexandercenter.com/jd/alexandertechniquejones.html.
3. https://twitter.com/dalailama/status/1082222631816818688?lang=en.
4. Tony Landau, "Primum Non Nocere: Latin for 'First, Do No Harm,'" *Forbes*, January 24, 2019, https://www.forbes.com/sites/impactpartners/2019/01/24 /primum-non-nocere-latin-for-first-do-no-harm/#dccf6e266014.
5. Fred Kofman, *Conscious Business: How to Build Value Through Values* (Boulder: Sounds True, 2006).
6. "Remembering Spiritual Masters Project: Bernie Glassman," Spirituality and Practice website, https://www.spiritualityandpractice.com/explorations/teachers /bernie-glassman/quotes.
7. Peter Economy, "17 Powerfully Inspiring Quotes from Southwest Airlines Founder Herb Kelleher, *Inc.*, January 4, 2019, https://www.inc.com/peter-economy/17 -powerfully-inspiring-quotes-from-southwest-airlines-founder-herb-kelleher.html.
8. Robert I. Sutton, "Are You a Jerk at Work?," *Greater Good* magazine, December 1, 2007, https://greatergood.berkeley.edu/article/item/are_you_jerk_work.
9. Robert I. Sutton, *The No Asshole Rule: Building a Civilized Workplace and Surviving One That Isn't* (New York: Business Plus, 2007).

10. https://www.southwestmag.com/herb-kelleher/.
11. Debankan Chattopadhyay, "The Customer Is Always Right. Hmm, Really?" CCE, June 6, 2017, http://www.cadcam-e.com/blogs/10/Customer_Is_Always _Right.aspx.
12. Philip Zimbardo, "Why the World Needs Heroes," *Europe's Journal of Psychology* 7(3), 402–7, reprinted at https://www.facebook.com/notes/heroic -imagination-project/why-the-world-needs-heroes/10150215562879229/.
13. Paul Stamets on Tim Ferriss Podcast, October 11, 2018, https://tim.blog /2018/10/11/paul-stamets/.
14. Theodore Roosevelt, "Address at the Prize Day Exercises at Groton School," Groton, Massachusetts, May 24, 1904.

ON BECOMING A HEALING LEADER

1. Brendan Byrnes, "An Interview with Jim Sinegal, CoFounder of Costco," The Motley Fool, July 31, 2013, https://www.fool.com/investing/general/2013/07/31 /an-interview-with-jim-sinegal-of-costco.aspx.
2. Perkins, *The World Is As You Dream It*.
3. S. Radhakrishnan, *The Bhagavadgita* (New York: Harper & Row, 1973).
4. As quoted in *Of Permanent Value: The Story of Warren Buffett* (Birmingham: AKPE, 2007) by Andrew Kilpatrick.
5. Michael J. Gelb, *How to Think Like Leonardo da Vinci: Seven Steps to Genius Every Day* (New York: Bantam Dell, 2004), 226.
6. Kahlil Gibran, "On Work," Katsandogz.com, http://www.katsandogz.com/on work.html.

EPILOGUE

1. "Franklin D. Roosevelt, Day by Day: October 1940," FDR Presidential Library, Marist University, http://www.fdrlibrary.marist.edu/daybyday/event/october -1940-10/.
2. Robert F. Kennedy, "Day of Affirmation Address," University of Capetown, Capetown, South Africa, June 6, 1966, from the John F. Kennedy Presidential Library and Museum, https://www.jfklibrary.org/learn/about-jfk/the-kennedy -family/robert-f-kennedy/robert-f-kennedy-speeches/day-of-affirmation -address-university-of-capetown-capetown-south-africa-june-6-1966.

INDEX

INDEX

INDEX

INDEX

INDEX

INDEX

INDEX